Peter Würfel

Physics of Solar Cells

Related Titles

Brendel, R.

Thin-Film Crystalline Silicon Solar Cells

Physics and Technology

2002

ISBN: 978-3-527-40376-9

Markvart, T. (ed.)

Solar Electricity

2000

ISBN: 978-0-471-98853-3

Goetzberger, A., Knobloch, J., Voss, B.

Crystalline Silicon Solar Cells

1998

ISBN: 978-0-471-97144-3

Peter Würfel

Physics of Solar Cells

From Basic Principles to Advanced Concepts

2nd, updated and expanded edition

Problems and Solutions by Uli Würfel

WILEY-VCH Verlag GmbH & Co. KGaA

The Authors

Prof. Dr. Peter Würfel
Universität Karlsruhe
Karlsruhe, Germany
peter.wuerfel@phys.uni-karlsruhe.de

Dr. Uli Würfel
Freiburger Materialforschungszentrum
Universität Freiburg
Freiburg, Germany
uli.wuerfel@fmf.uni-freiburg.de

Cover picture

The sun rising over the earth
Video by John Feather
www.serpentsheart.com

Library of Congress Card No.: applied for

British Library Cataloguing-in-Publication Data
A catalogue record for this book is available from the British Library.

Bibliographic information published by the Deutsche Nationalbibliothek
The Deutsche Nationalbibliothek lists this publication in the Deutsche Nationalbibliografie; detailed bibliographic data are available on the Internet at ⟨http://dnb.d-nb.de⟩.

Composition Laserwords Private Ltd., Chennai, India

Printed on acid-free paper

ISBN: 978-3-527-40857-3

Contents

Physics of Solar Cells: From Basic Principles to Advanced Concepts. Peter Würfel

ISBN: 978-3-527-40857-3

List of Symbols

h, $\hbar = h/(2\pi)$	Planck's constant	eVs
$\hbar\omega$	photon energy	eV
$a(\hbar\omega)$	absorptance	
$r(\hbar\omega)$	reflectance	
$t(\hbar\omega)$	transmittance	
$\varepsilon(\hbar\omega) = a(\hbar\omega)$	emittance	
$\alpha(\hbar\omega)$	absorption coefficient	cm^{-1}
k	Boltzmann's constant	$eV\ K^{-1}$
σ	Stefan–Boltzmann constant	$W\ m^{-2}K^{-4}$
T	temperature	K
n_j	concentration of particle type j	cm^{-3}
e	electron	
h	hole	
γ	photon	
Γ	phonon	
n_e, n_h	concentration of electrons, holes	cm^{-3}
n_i	intrinsic concentration of electrons and holes	cm^{-3}
N_C, N_V	effective density of states in conduction band, valence band	cm^{-3}
ε_e, ε_h	energy of an electron, hole	eV
ε_C	energy of an electron at the conduction band minimum	eV
ε_V	energy of an electron at the valence band maximum	eV
μ_j	chemical potential of particle type j	eV
η_j	electrochemical potential of particle type j	eV
χ_e	electron affinity	eV
φ	electrical potential	V
e	elementary charge	As
ϵ_0	dielectric permittivity of free space	$As\ (V\ m)^{-1}$
ϵ	relative dielectric permittivity	

Physics of Solar Cells: From Basic Principles to Advanced Concepts. Peter Würfel

ISBN: 978-3-527-40857-3

V	voltage = $[\eta_e(x_1) - \eta_e(x_2)]/e$	V
ε_{FC}	Fermi energy for electron distribution in conduction band	eV
ε_{FV}	Fermi energy for electron distribution in valence band	eV
m_e^*, m_h^*	effective mass of electrons, holes	g
b_e, b_h	mobility of electrons, holes	$cm^2\ (Vs)^{-1}$
D_e, D_h	diffusion coefficient of electrons, holes	$cm^2\ s^{-1}$
τ_e, τ_h	recombination life time of electrons, holes	s
R_e, R_h	recombination rate of electrons, holes	$cm^{-3}s^{-1}$
G_e, G_h	generation rate of electrons, holes	$cm^{-3}s^{-1}$
σ_e, σ_h	cross-section for the capture of an electron, hole by an impurity	cm^2
j_j	current density of particles of type j	$(cm^2s)^{-1}$
j_Q	charge current density	$A\ cm^{-2}$

Preface

Mankind needs energy for a living. Besides the energy in our food necessary to sustain our body and its functions (100 W), 30 times more energy is used on average to make our life more comfortable. Electrical energy is one of the most useful forms of energy, since it can be used for almost everything. All life on earth is based on solar energy following the invention of photosynthesis by the algae. Producing electrical energy through photovoltaic energy conversion by solar cells is the human counterpart. For the first time in history, mankind is able to produce a high quality energy form from solar energy directly, without the need of the plants. Since any sustainable, i.e. long term energy supply must be based on solar energy, photovoltaic energy conversion will become indispensable in the future.

This book provides a fundamental understanding of the functioning of solar cells. The discussion of the principles is as general as possible to provide the basis for present technology and future developments as well. Energy conversion in solar cells is shown to consist of two steps. The first is the absorption of solar radiation and the production of chemical energy. This process takes place in every semiconductor. The second step is the transformation into electrical energy by generating current and voltage. This requires structures and forces to drive the electrons and holes, produced by the incident light, through the solar cell as an electric current. These forces and the structures which enable a directional charge transport are derived in detail. In the process it is shown that the electric field present in a pn junction in the dark, usually considered a prerequisite for the operation of a solar cell, is in fact more an accompanying phenomenon of a structure required for other reasons and not an essential property of a solar cell. The structure of a solar cell is much better represented by a semiconducting absorber in which the conversion of solar heat into chemical energy takes place and by two semi-permeable membranes which at one terminal transmit electrons and block holes and at the second terminal transmit holes and block electrons. The book attempts to develop the physical principles underlying the function of a solar cell as understandably and at the same time as completely as possible. With very few exceptions, all physical relationships are derived and explained in

Physics of Solar Cells: From Basic Principles to Advanced Concepts. Peter Würfel

ISBN: 978-3-527-40857-3

examples. This will provide the nonphysicists particularly with the background for a thorough understanding.

Emphasis is placed on a thermodynamic approach that is largely independent of existing solar cell structures. This allows a general determination of the efficiency limits for the conversion of solar heat radiation into electrical energy and also demonstrates the potential and the limits for improvement for present-day solar cells. We follow a route first taken by W. Shockley and H. J. Queisser.[1)]

This book is the result of a series of lectures dealing with the physics of solar cells. I am grateful to the many students who called my attention to errors and suggested improvements. The material presented here, which differs from the usual treatment of solar cells relying on an electric field for a driving force, is the result of several years of collaboration with my teacher, W. Ruppel.

In some respects this book is more rigorous than is customary in semiconductor device physics and in solar cell physics in particular. The most obvious is that identical physical quantities will be represented by identical symbols. Current densities will be represented by j and the quantity that is transported by the current is defined by its index, as in j_Q for the density of a charge current or j_e for the density of a current of electrons. In adhering to this principle, all particle concentrations are given the symbol n, with n_e representing the concentration of electrons, n_h the concentration of holes and n_γ the concentration of photons. I hope that those who are used to n and p for electron and hole concentrations do not find it too difficult to adapt to a more logical notation.

The driving force for a transition from exhausting energy reserves, as we presently do, to using renewable energies, is not the exhaustion of the reserves themselves, although oil and gas reserves will not last for more than one hundred years. The exhaustion does not bother most of us, since it will occur well beyond our own lifetime. We would certainly care a lot more, if we were to live for 500 years and would have to face the consequences of our present energy use ourselves. The driving force for the transition to renewable energies is rather the harmful effect which the byproducts of using fossil and nuclear energy have on our environment. Since this is the most effective incentive for using solar energy, we start by discussing the consequences of our present energy economy and its effect on the climate. The potential of a solar energy economy to eliminate these problems fully justifies the most intensive efforts to develop and improve the photovoltaic technology for which this book tries to provide the foundation.

Peter Würfel

1) W. Shockley, H. J. Queisser, *J. Appl. Phys.* **32**, (1961), 510.

1
Problems of the Energy Economy

The energy economy of nearly all and, in particular, of the industrialized countries is based on the use of stored energy, mainly fossil energy in the form of coal, oil, and natural gas, as well as nuclear energy in the form of the uranium isotope ^{235}U. Two problems arise when we use our reserves to satisfy our energy needs. A source of energy can continue only until it is depleted. Well before this time, that is, right now at the latest, we have to consider how life will continue after this source of energy is gone and we must begin to develop alternatives. Furthermore, unpleasant side effects accompany the consumption of the energy source. Materials long buried under the surface of the Earth are released and find their way into air, water, and into our food. Up to now, the disadvantages are hardly perceptible, but they will lead to difficulties for future generations. In this chapter, we estimate the size of the fossil energy resources, which, to be precise, are composed not only of fossil energy carriers but also of the oxygen in the air that burns with them. In addition, we examine the cause of the greenhouse effect, which is a practically unavoidable consequence of burning fossil fuels.

1.1
Energy Economy

The amount of chemical energy stored in fossil energy carriers is measured in energy units, some more and some less practical. The most fundamental unit is the joule, abbreviated J, which is, however, a rather small unit representing the amount of energy needed to heat 1 g of water by a quarter of a degree or the amount of energy that a hair drier with a power of 1 kW consumes in 1 ms. A more practical unit is the kilo Watt hour (kWh), which is 3.6×10^6 J. 1 kWh is the energy contained in 100 g of chocolate. The only problem with this unit is that it is derived from the watt, the unit for power, which is energy per time. This makes energy equal to power times time. This awkwardness leads to a lot of mistakes in the nonscience press such as kilowatt per hour for power, because most people mistake kilowatt to be the unit for energy, which they perceive as the more basic quantity. The energy of fossil fuels is often given in barrels of oil equivalents or in (metric) tons of coal equivalents (t coal equiv.).

Physics of Solar Cells: From Basic Principles to Advanced Concepts. Peter Würfel

ISBN: 978-3-527-40857-3

The following relations apply:

$$1\ \text{kWh} = 3.6 \times 10^6\ \text{J} = 1\ \text{kWh}$$
$$1\ \text{t coal equiv.} = 29 \times 10^9\ \text{J} = 8200\ \text{kWh}$$
$$1\ \text{kg oil} = 1.4\ \text{kg coal equiv.} = 12.0\ \text{kWh}$$
$$1\ \text{m}^3\ \text{gas} = 1.1\ \text{kg coal equiv.} = 9.0\ \text{kWh}$$
$$1\ \text{barrel oil} = 195\ \text{kg coal equiv.} = 1670\ \text{kWh}$$

The consumption of chemical energy per time is an energy current (power) taken from the energy reserves. Thus, the consumption of one ton of coal per year, averaged over one year amounts to an energy current of

$$1\ \text{t coal equiv./a} = 8200\ \text{kWh/a} = 0.94\ \text{kW}$$

We look at Germany as an example of a densely populated industrialized country. Table 1.1 shows the consumption of primary energy in Germany in the year 2002, with a population of 82.5×10^6. These figures contain a consumption of electrical energy per year of

$$570\ \text{TWh/a} = 65\ \text{GW} \Longrightarrow 0.79\ \text{kW/person}$$

The energy consumption per capita in Germany of 5.98 kW is very high compared with the energy current of 2000 kcal/d = 100 W = 0.1 kW taken up by human beings in the form of food, representing the minimum requirement for sustaining life.

Table 1.2 shows the consumption of primary energy in the world in 2002, with a population of 6×10^9. This energy consumption is supplied from the available reserves of energy with the exception of hydro, wind, and biomass. The current remaining reserves of energy are shown in Table 1.3. This is the amount of energy that is estimated to be recoverable economically with present-day techniques at current prices. The actual reserves may be up to 10 times as large, about 10×10^{12} t coal equiv.

Table 1.1 Primary energy consumption in Germany in 2002.

Type	Consumption (10^6 t coal equiv./a)	Per capita consumption (kW/person)
Oil	185	2.24
Gas	107	1.30
Coal	122	1.48
Nuclear energy	62	0.75
Others	17	0.21
Total	494	5.98

Table 1.2 World primary energy consumption in 2002.

Type	Consumption (10^9 t coal equiv./a)	Per capita consumption (kW/person)
Oil	4.93	0.82
Gas	3.19	0.53
Coal	3.36	0.56
Nuclear energy	0.86	0.14
Others	0.86	0.14
Total	13.2	2.19

Table 1.3 The world's remaining energy reserves.

Type	Reserves in 10^9 t coal equiv.
Oil	210
Gas	170
Coal	660
Total	1040

The global energy consumption of 13.2×10^9 t coal equiv. per year appears to be very small when compared with the continuous energy current from the Sun of

$$1.7 \times 10^{17}\ \text{W} = 1.5 \times 10^{18}\ \text{kWh/a} = 1.8 \times 10^{14}\ \text{t coal equiv./a}$$

that radiates toward the Earth.

In densely populated regions such as Germany, however, the balance is not so favorable if we restrict ourselves to the natural processes of photosynthesis for the conversion of solar energy into other useful forms of energy. The mean annual energy current that the Sun radiates onto Germany, with an area of 0.36×10^6 km^2, is about 3.6×10^{14} kWh/a $= 4.3 \times 10^{10}$ t coal equiv./a. Photosynthesis, when averaged over all plants, has an efficiency of about 1% and produces around 400×10^6 t coal equiv./a from the energy of the Sun. This is insufficient to cover the requirements of primary energy of 494×10^6 t coal equiv./a for Germany. What is even more important is that it also shows that over the entire area of Germany plants are not able to reproduce, by photosynthesis, the oxygen that is consumed in the combustion of gas, oil and coal. And this does not even take into consideration that the biomass produced in the process is not stored but decays, which again consumes the oxygen produced by photosynthesis. This estimate also shows that solar energy can cover the energy requirements of Germany over its area only if a substantially higher efficiency for the conversion process than that of photosynthesis can be achieved. The fact that no shortage in the supply of oxygen will result in the foreseeable future is owed to the wind, which brings oxygen from areas with lower consumption. Nevertheless, well before we run out of oxygen we will be made aware of an increase in the combustion product CO_2.

1.2 Estimate of the Maximum Reserves of Fossil Energy

For this estimate [1] we assume that neither free carbon nor free oxygen was present on the Earth before the beginning of organic life. The fact that carbon and oxygen react quickly at the high temperatures prevailing during this stage of the Earth's history, both with each other to form CO_2 and also with a number of other elements to form carbides and oxides, is a strong argument in support of this assumption. Since there are elementary metals on the surface of the Earth even today, although only in small amounts, it must be assumed that neither free carbon nor free oxygen was available to react.

The free oxygen found in the atmosphere today can therefore only be the result of photosynthesis occurring at a later time. The present-day amount of oxygen in the atmosphere thus allows us to estimate the size of the carbon reserves stored in the products of photosynthesis.

During photosynthesis, water and carbon dioxide combine to form carbohydrates, which build up according to the reaction

$$n \times (H_2O + CO_2) \Longrightarrow n \times CH_2O + n \times O_2$$

A typical product of photosynthesis is glucose: $C_6H_{12}O_6 \equiv 6 \times CH_2O$. For this compound and also for most other carbohydrates, the ratio of free oxygen produced by photosynthesis to carbon stored in the carbohydrates is

$$1\ \text{mol}\ O_2 \Longrightarrow 1\ \text{mol C} \quad \text{or}$$

$$32\ \text{g}\ O_2 \Longrightarrow 12\ \text{g C}$$

The mass of the stored carbon m_C can therefore be found from the mass of free oxygen m_{O_2}:

$$m_C = \frac{12}{32} m_{O_2}$$

The greatest proportion of the oxygen resulting from photosynthesis is found in the atmosphere and, to a lesser extent, dissolved in the water of the oceans. The fraction in the atmosphere is sufficiently large to be taken as the basis for an estimate.

From the pressure $p_E = 1\ \text{bar} = 10\ \text{Ncm}^{-2}$ on the surface of the Earth resulting from the air surrounding our planet, we can calculate the mass of air from the relationship $m_{air} \times g = p_E \times$ area:

$$m_{\text{air}}/\text{area} = p_E/g = \frac{10\ \text{N cm}^{-2}}{10\ \text{m s}^{-2}} = 1\ \text{kg cm}^{-2}$$

Multiplying by the surface of the Earth gives the total mass of air

$$m_{\text{air}} = 1\ \text{kg cm}^{-2} \times 4\pi R^2_{\text{Earth}} = 5 \times 10^{15}\ \text{tons of air}$$

Since air consists of 80% N_2 and 20% O_2 (making no distinction between volume percent and weight percent), the mass of oxygen is: $m_{O_2} = 10^{15}$ t O_2. The maximum amount of carbon produced in photosynthesis and now present in deposits on the Earth is therefore:

$$m_C = \frac{12}{32} m_{O_2} = 375 \times 10^{12}\ \text{tons of carbon}$$

Up to now 10.4×10^{12} t coal equiv. has been found.

Thus, there is reason to hope that the reserves of fossil energy will continue to grow as a result of continued exploration. In fact, in recent years the known reserves have grown continuously because more has been found than was consumed. There are rumors that very large reserves of methane hydrate can be found in moderate depths on the ocean floor. This compound dissociates into methane and water when it is heated or taken out of the ocean. The prospects of possibly large reserves, however, must not distract our attention from the urgency of restricting the mining of these reserves. If we actually use up the entire reserves of carbon for our energy requirements, we will in fact reverse the photosynthesis of millions of years and in doing so eliminate all our oxygen. Even if more than the estimated 375×10^{12} tons of carbon should exist, we cannot burn more than 375×10^{12} tons of carbon because of the limited amount of oxygen.

If we examine oil and gas consumption as an example for the consumption of fossil energy reserves over a long period of time, for example, since the birth of Christ, we obtain a frightening picture (Figure 1.1). Up to the beginning of the twentieth century, the consumption of reserves was practically negligible. From then, it has been rising exponentially and will reach a maximum value in one or two decades. Consumption will then fall off again as the reserves are gradually used up. The maximum consumption is expected when the reserves have fallen off to one half of

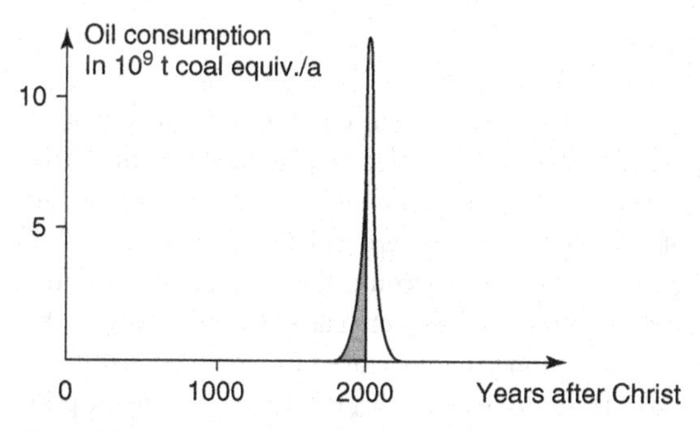

Figure 1.1 Annual consumption of oil. The area under the curve gives the estimated total oil reserves.

their original levels. The reserves that have accumulated over millions of years will then literally go up in smoke over a period of only about 100 years. Here, the elimination of energy reserves is the lesser problem. Much worse will be the alteration of the atmosphere as a result of the products of combustion. These effects will last for a long time. Even if later generations change over to supplying energy from regenerative sources, they will still suffer from the heritage that we have left them.

1.3 The Greenhouse Effect

CO_2 is produced in the combustion of fossil energy carriers. The increase in the concentration of carbon dioxide in our atmosphere will have serious consequences on our climate. Currently, the atmosphere contains a fraction of 0.03% of CO_2. This corresponds to 2.3×10^{12} t of CO_2.

1.3.1 Combustion

Pure carbon is consumed according to the reaction $C + O_2 \Rightarrow CO_2$. Accordingly $12\,g\,C + 32\,g\,O_2 \Rightarrow 44\,g\,CO_2$. The mass of CO_2 produced by combustion is given by the mass of carbon consumed according to $m_{CO_2} = 44/12 m_C$. The combustion of 1 t of carbon thus results in 3.7 t of CO_2.

For different compounds of carbon other relationships apply:

- Carbohydrates:

$$30\,g\,CH_2O + 32\,g\,O_2 \Longrightarrow 18\,g\,H_2O + 44\,g\,CO_2 \tag{1.1}$$

  resulting in $m_{CO_2} = 1.47\, m_{CH_2O}$. This is the chemical reaction for the combustion of food in the human body.

- Methane (main component of natural gas):

$$16\,g\,CH_4 + 64\,g\,O_2 \Longrightarrow 36\,g\,H_2O + 44\,g\,CO_2$$

  resulting in $m_{CO_2} = 2.75\, m_{CH_4}$.

The present global consumption of the 10^{10} t coal equiv./a produces globally $\Rightarrow 2.2 \times 10^{10}$ t of CO_2 per year. Half of this is dissolved in the water of the oceans and half remains in the atmosphere. If the annual energy consumption does *not* continue to rise, the amount of CO_2 in the atmosphere will double only after about 200 years. However, it is necessary to take into account that energy consumption continues to increase. Currently the growth of 1% per year is relatively low. In the developing countries, energy consumption even decreased in 1999, because of the inability of these countries to pay for more energy. If global energy consumption continues to increase at about 1% per year the CO_2 concentration in the atmosphere will have doubled after about 100 years. This increase is less the result of a per capita increase in energy consumption than that of the increasing global population. The increasing CO_2 concentration in the atmosphere will have consequences on the temperature of the Earth.

1.3.2
The Temperature of the Earth

The temperature of the Earth is stationary, i.e. constant in time if the energy current absorbed from the Sun and the energy current emitted by the Earth are in balance. We want to estimate the temperature of the Earth in this steady state condition. For this purpose we make use of radiation laws, not derived until the following chapter.

The energy current density from the Sun at the position of the Earth (but outside the Earth's atmosphere) is

$$j_{E,\,\text{Sun}} = 1.3\ \text{kW m}^{-2}$$

For the case of complete absorption, the energy current absorbed by the entire Earth is the energy current incident on the projected area of the Earth:

$$I_{E,\,\text{abs}} = \pi R_e^2 j_{E,\,\text{Sun}} \quad \text{with} \quad R_e = 6370\ \text{km} \quad \text{(radius of the Earth)}$$

According to the Stefan–Boltzmann radiation law, the energy current density emitted by the Earth into space is given by

$$j_{E,\,\text{Earth}} = \sigma T_e^4 \quad \text{where} \quad \sigma = 5.67 \times 10^{-8}\ \text{W m}^{-2}\text{K}^{-4}$$

is the Stefan–Boltzmann constant. The energy current emitted by the entire Earth is

$$I_{E,\,\text{emit}} = 4\pi R_e^2 \sigma T_e^4$$

From the steady state condition $I_{E,\,\text{abs}} = I_{E,\,\text{emit}}$, it follows that the estimated mean temperature of the Earth is $T_e = 275$ K.

The mean temperature of the Earth is in fact around 288 K. The approximate agreement is, however, only coincidental. Taking into account that about 30% of the incident solar radiation is reflected back into space by the atmosphere of the Earth and thus only about 70% (1 kW m^{-2}) reaches the surface of the Earth, a temperature of 258 K then results. The actual temperature of the Earth is in fact greater, because the radiation emitted by the Earth is partly absorbed in the atmosphere. The atmosphere then becomes warmer and emits heat back to the Earth. The same effect occurs in greenhouses, where the glass covering absorbs the thermal radiation emitted by the greenhouse and emits some of it back into the greenhouse.

We can understand the greenhouse effect of the atmosphere using a simple model. Owing to a temperature of 6000 K of the Sun, the solar radiation spectrum (expressed as energy current per wavelength) has a maximum at about 0.5 μm, at which the atmosphere is transparent. As a result of the lower temperature of the Earth, the emission from the Earth has its maximum at about 10 μm (in the infrared region). All triatomic molecules, including CO_2, are good absorbers in the infrared region. Consequently, while most of the incident solar radiation reaches the surface of the Earth, a great part of the energy emitted from the surface is absorbed in the atmosphere. This causes warming of the atmosphere, which in turn leads to heat emission back to the Earth. The temperature of the Earth is at maximum when the radiation from the surface of the Earth is completely absorbed by the atmosphere, a situation that will be faced if the atmospheric concentration of CO_2 continues to rise.

In our model, we assume that the Earth's surface absorbs all radiation incident upon it from the Sun and the atmosphere. In the steady state, it must also emit the same energy current. All energy emitted in the infrared region is assumed to be absorbed by the atmosphere. This leads to the condition of Figure 1.2:

$$I_{E,\,\mathrm{Earth}} = I_{E,\,\mathrm{Sun}} + \tfrac{1}{2} I_{E,\,\mathrm{atm}}$$

Since radiation emitted by the Earth's surface is fully absorbed by the atmosphere, the solar energy current incident on the surface of the Earth can only be emitted into space by emission from the atmosphere. This leads to the following conditions:

$$\tfrac{1}{2} I_{E,\,\mathrm{atm}} = I_{E,\,\mathrm{Sun}} \quad \text{and} \quad I_{E,\,\mathrm{Earth}} = 2 I_{E,\,\mathrm{Sun}}$$

It then follows that

$$I_{E,\,\mathrm{Earth}} = 4\pi R_\mathrm{e}^2 \sigma T_{\mathrm{e,\,greenhouse}}^4 = 2\pi R_\mathrm{e}^2 \times 1.3\ \mathrm{kW\ m^{-2}}$$

This yields a temperature of

$$T_{\mathrm{e,\,greenhouse}} = \sqrt[4]{2}\ T_\mathrm{e} = 1.19 \times 275\ \mathrm{K} = 327\ \mathrm{K} = 54\,^\circ\mathrm{C}$$

With this mean temperature, the Earth would be virtually uninhabitable.

What makes the problem of the increased absorption of infrared radiation in the atmosphere due to human influences even worse is that only one half of the present greenhouse effect is caused by the increasing CO_2 concentration. The other half results from methane, fluorinated hydrocarbons, and nitrogen oxides.

To estimate the greenhouse effect, we have treated the atmosphere as a fire screen, allowing the solar radiation to pass through but absorbing the radiation from the Earth. In view of the fact that the temperature of the atmosphere is not uniform, a better description would be in the form of several fire screens placed behind each other. Extending the fire-screen model to n fire screens, the temperature of the Earth's surface becomes

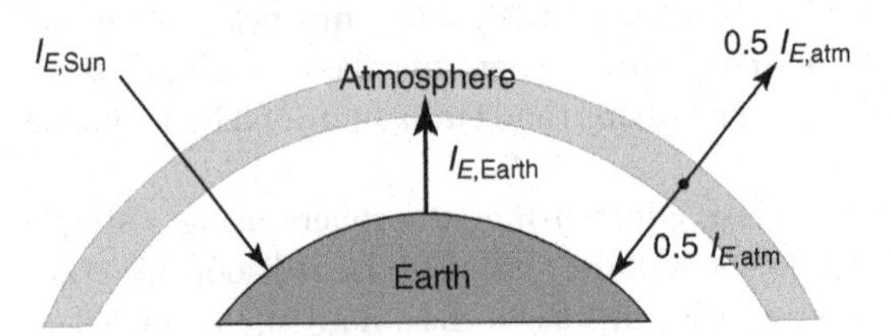

Figure 1.2 Balance of the absorbed and emitted energy currents on the surface of the Earth.

$$T_{\mathrm{e,greenhouse}} = T_{\mathrm{e}}\sqrt[4]{n+1}$$

For large values of n, the temperature of the Earth could become arbitrarily large and, in particular, even greater than the temperature of the Sun. This result is of course not correct, because at such high temperatures an essential condition of the model is no longer satisfied. If the temperature of the Earth were equal to that of the Sun, the spectrum emitted from the Earth would be identical with the solar spectrum. It would then no longer be possible for the solar radiation to pass through the atmosphere, while the radiation emitted from the Earth would be absorbed, as assumed for the fire-screen model.

As long as the requirements are satisfied, with more than one fire screen a higher temperature results. The conditions prevailing on Venus clearly show this. The distance of Venus from the Sun is 0.723 times the distance of the Earth from the Sun. The energy current density of the solar radiation at the position of Venus is therefore twice that reaching the Earth. The atmosphere of Venus is composed almost entirely of CO_2. Treating the atmosphere of Venus as a single fire screen yields a temperature of 116 °C for Venus, a factor of $\sqrt[4]{2}$ greater than that of the Earth. The actual temperature of Venus is, however, 475 °C.

The number of fire screens required to describe the atmosphere depends on the mean free path of the emitted radiation after which it is absorbed in the atmosphere. This determines the spacing of the fire screens in the model. Because of the high density of Venus' atmosphere, with a pressure of 90 bar, this mean free path is much shorter on Venus than on Earth.

Fortunately, we do not have to worry about temperatures on the Earth reaching those found on Venus. Even if the entire supply of oxygen on the Earth were consumed, resulting in a CO_2 pressure of about 0.2 bar, the temperature of the Earth could never reach the temperature of Venus. Nevertheless, serious alterations are already occurring even with much lower temperature increases, which are not only possible but in fact very probable on Earth. We also have to consider that there are numerous feedback effects. The most dangerous would be the release of large quantities of methane, an even more effective greenhouse gas than CO_2, when methane hydrate melts in a warmer ocean.

1.4 Problems

1.1 Describe in a few sentences the greenhouse effect

- **(a)** in a greenhouse for crops
- **(b)** in the Earth's atmosphere.

1.2 Why does burning methane lead to a higher greenhouse activity than burning (the same mass of) carbohydrates?

1.3 How long may a computer (50 W) be operated with an amount of energy of 1 kg of coal equiv.?

1.4 Assume the global human energy consumption to be 15×10^9 t coal equiv./a by now and the Earth's remaining consumable energy reserves to be 2×10^{12} t coal equiv. How long will these reserves last with the energy consumption

(a) increasing linearly, doubling in 35 years?

(b) increasing exponentially, doubling in 40 years?

1.5 Calculate the amount of methane present on Earth if all the oxygen in the atmosphere would have resulted from the production of methane from CO_2. Consider a total mass of oxygen of $m_{O_2} = 10^{15}$ t.

1.6 Why has Venus such a high surface temperature? Venus is the brightest object in the sky (except for the Sun and Moon) due to its high reflectance. This so-called albedo of Venus is $r_v = 0.7$. Calculate the intensity reaching the surface of Venus. Knowing the intensity and the surface temperature (475 °C), determine the number of fire screens according to the model in Section 1.3.2.

2
Photons

Photons are particles of light. Photons to which the human eye reacts, that is, those that we see, have energies $\hbar\omega$ between 1.5 and 3 eV. They always move with the velocity of light, in a vacuum with a velocity of $c_0 = 3 \times 10^8$ m s^{-1}, and in a medium with refractive index n with a velocity of $c = c_0/n$. The fact that we can also describe light as an electromagnetic wave is in no way contradictory. The square of the field strength of the electromagnetic wave describes the location of the photons. In contrast to the behavior familiar from shot pellets, photons obey the laws of quantum mechanics and not those of the more common Newtonian mechanics. The differences from Newtonian mechanics become apparent only for particles of very low energy (including the mass). Since it is difficult to visualize particles that do not move along a straight line, in a dualistic approach the wave property is generally invoked to describe diffraction and interference phenomena and the particle property is used to describe the quantumlike transport of energy, as in this book.

2.1
Black-body Radiation

A *black body* is defined as a body that completely absorbs radiation of all photon energies $\hbar\omega$. Its absorptance is

$$a(\hbar\omega) = 1 \tag{2.1}$$

A possible example of a black body is a tiny hole in a cavity. Photons incident on the hole are not reflected, but are absorbed inside the cavity. The hole is described as black, because it appears black when the inner part of the cavity is at a low temperature and very few photons are emitted by the walls of the cavity. When the inner part is at a high temperature, the cavity is like a furnace. We then see a very bright hole, through which we can view the interior of the furnace. Nevertheless, regardless of whether it emits radiation, the hole absorbs all incident radiation and is therefore still described as a black body.

The Sun is a black body, as is every body that is sufficiently thick and nonreflecting.

Physics of Solar Cells: From Basic Principles to Advanced Concepts. Peter Würfel

ISBN: 978-3-527-40857-3

2.1.1
Photon Density n_γ in a Cavity (Planck's Law of Radiation)

Planck's law of radiation describes the photon density $dn_\gamma(\hbar\omega)$ in a cavity for photon energies between $\hbar\omega$ and $\hbar\omega + d\hbar\omega$. As always for the definition of particle densities, we first determine the density D of states for these particles. The particle density is then found by applying an occupation or distribution function f. For photons, this takes the form

$$dn_\gamma(\hbar\omega) = D_\gamma(\hbar\omega) f_\gamma(\hbar\omega)\, d\hbar\omega \tag{2.2}$$

Here $D_\gamma(\hbar\omega)$ is the density (per volume and per energy interval $d\hbar\omega$) of the states that the photons can occupy and $f_\gamma(\hbar\omega)$ is the probability of occupation or the distribution function that determines the distribution of photons over the states as a function of the energy $\hbar\omega$.

Distribution Function $f_\gamma(\hbar\omega)$ for Photons

Because of their integral spin, the Bose–Einstein distribution (Figure 2.1), which describes the probability for the occupation of states with the energy $\hbar\omega$, applies for the photons:

$$f_\gamma(\hbar\omega) = \frac{1}{\exp\left[(\hbar\omega - \mu_\gamma)/kT\right] - 1} \tag{2.3}$$

Here, and always in the combination kT, k denotes the Boltzmann constant $k = 8.617 \times 10^{-5}$ eV K^{-1}, and μ_γ is the chemical potential of the photons, which has the value $\mu_\gamma = 0$ for solar radiation and in general for thermal radiation. Since the distribution function $f_\gamma(\hbar\omega)$ depends only on the energy $\hbar\omega$ of the photons, we must also know the density of the states $D_\gamma(\hbar\omega)$ as a function of the energy.

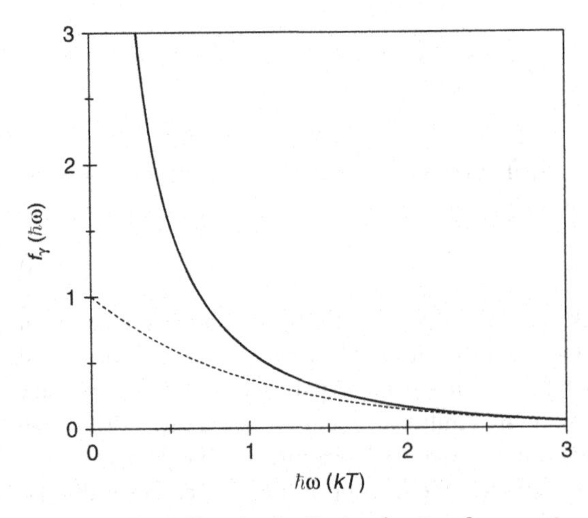

Figure 2.1 Bose–Einstein distribution function for $\mu_\gamma = 0$. The dashed line represents the Boltzmann distribution $\exp(-\hbar\omega/kT)$, a good approximation for $\hbar\omega > 2kT$.

Density of States $D_\gamma(\hbar\omega)$ for Photons

Two particles can be distinguished only if they are in different states. Two states are different when their location x and momentum p differ by more than the minimum of the uncertainty Δx, Δp of one of these states. For one dimension, this Heisenberg uncertainty principle has the form

$$\Delta x \Delta p_x \geq h \tag{2.4}$$

Particles that cannot be distinguished as a result of the uncertainty principle are in the same state. In geometrical and momentum space, a state therefore has the "phase space" volume

$$\Delta x \Delta p_x \Delta y \Delta p_y \Delta z \Delta p_z = h^3$$

Since the photons are not localized in the cavity, the uncertainty in their position must be taken as the entire volume of the cavity, extending in the x, y, and z directions by L_x, L_y, and L_z. Thus,

$$\Delta x = L_x \quad \text{and consequently} \quad \Delta p_x = h/L_x$$

Possible values for the x component of the momentum lie in intervals with a width of h/L_x with

$$p_x = 0, \ \pm\frac{h}{L_x}, \ \pm\frac{2h}{L_x}, \ldots$$

The relationships for Δp_y and Δp_z are analogous. States having the volume $V = L_x L_y L_z$ in geometrical space then have the volume $\Delta p_x \Delta p_y \Delta p_z = \Delta p^3 = h^3\ V^{-1}$ in momentum space. Figure 2.2 illustrates the volume of the states and their homogeneous distribution in momentum space. This result is based solely on the Heisenberg uncertainty principle and does not contain any specific particle properties and for this reason it is not only valid for photons but also for electrons, which will be treated later.

A property specific to light is its occurrence in two directions of polarization normal to each other with independent intensities. It is therefore necessary to consider two independent types of photons (with opposite spins). Each volume $\Delta p^3 = h^3\ V^{-1}$ in momentum space therefore contains two photon states, one for each type of photon.

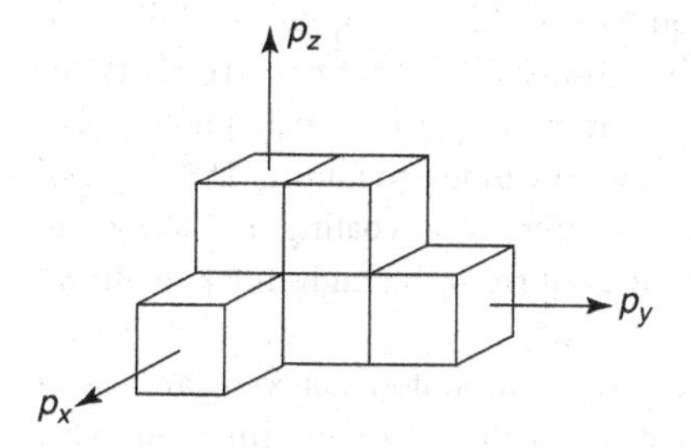

Figure 2.2 Volumes of the states in momentum space.

Since the distribution function $f_\gamma(\hbar\omega)$ for the photons depends on energy and not on momentum, we need the density of photon states per energy interval, rather than the density of states in momentum space. For this we need the relationship between momentum p and energy $\hbar\omega$

$$\hbar\omega = c\,|p|$$

which is a specific property of photons.

All states in which the photons have energies $\hbar\omega' < \hbar\omega$ and thus $|p'| < |p|$ lie within a sphere of radius $|p| = \hbar\omega/c$ in momentum space. The number of such states is

$$N_\gamma = \frac{2 \times (4\pi/3)p^3}{(h^3/V)} = \frac{(8\pi/3)V(\hbar\omega)^3}{h^3c^3} \tag{2.5}$$

From Equation 2.5, we obtain the density of states per volume and energy interval as the increase $\mathrm{d}N_\gamma$ resulting from an increase in the energy of $\mathrm{d}\hbar\omega$.

$$D_\gamma(\hbar\omega) = \frac{1}{V} \times \frac{\mathrm{d}N_\gamma(\hbar\omega)}{\mathrm{d}\hbar\omega} = \frac{(\hbar\omega)^2}{\pi^2\hbar^3c^3} \tag{2.6}$$

Here $\hbar = h/2\pi$. The velocity c of the photons is their velocity in the cavity. If the cavity is empty, the photons have the velocity of light in vacuum c_0. With increasing index of refraction the velocity c_0/n of the photons decreases, resulting in an increasing density of states for the photons.

A problem results when photons pass from one medium to another with a lower index of refraction, i.e. from a higher to a lower density of states, such as from glass to air. To accommodate all photons in the low-index material with fewer photon states, the occupation function f_γ would have to increase. This, however, is not allowed by the second principle of thermodynamics, as will be shown later, since it would result in annihilation of entropy. For the same value of the distribution function $f_\gamma(\hbar\omega)$, to keep the entropy constant, only part of the photons can be accommodated in the low-index material. Nature chooses to totally reflect back into the high-index material the photons that cannot be accommodated in the fewer states of the medium with the lower index of refraction. Exchange of photons between two media is restricted to the same number of states in both media in such a way that all states in a restricted solid angle in the high-index medium communicate with all states in an unrestricted solid angle in the low-index medium. Snell's law of refraction is another consequence of this principle. In addition to total internal reflection, a discontinuity of the refractive index at an interface causes reflection even for photons within the solid angle range in which photons are exchanged, e.g. all photons incident from the low-index medium. This type of reflection, however, can be eliminated by an antireflection coating, in contrast to the total internal reflection of the photons incident from the high-index medium, but with momenta outside the restricted solid angle.

The density of states (Equation 2.6) accounts for all photon states, regardless of their direction of motion, which is the direction of their momentum p. At every location within the cavity, the motion of the photons is in all directions, i.e. their motion is isotropic. Of interest to us here are the photons that are able to leave the cavity through the hole. At any location within the cavity, these are the photons moving toward the hole, i.e. having a momentum vector pointing in this direction. Since the photons do not collide with each other and maintain their direction, we can find these photons by determining their density per solid angle at every location inside the cavity and multiply the density per solid angle by the solid angle element $d\Omega$ subtended by the hole. Imagine a sphere with a radius R given by the distance to the hole, around the location of interest of the photon density per solid angle $d\Omega$. This solid angle element $d\Omega$ includes all photons passing through the surface element dO of the sphere, which defines the hole. The solid angle is defined as $d\Omega = dO/R^2$.

The maximum value of Ω is 4π, for a solid angle, which includes all directions. For an isotropic distribution of states in momentum space, the density of states per solid angle is

$$D_{\gamma,\Omega}(\hbar\omega) = \frac{D_\gamma(\hbar\omega)}{4\pi}$$

Energy Distribution of Photons

The density of states and the distribution function lead us to Planck's law of radiation, which describes the number of photons per volume and per photon energy interval $d\hbar\omega$ in the solid angle element $d\Omega$.

$$\frac{dn_\gamma(\hbar\omega)}{d\hbar\omega} = D_{\gamma,\Omega}(\hbar\omega) f_\gamma(\hbar\omega)\, d\Omega = \frac{(\hbar\omega)^2\, d\Omega}{4\pi^3\hbar^3(c_0/n)^3} \frac{1}{\exp(\hbar\omega/kT) - 1} \tag{2.7}$$

The photon density per photon energy interval $dn_\gamma/d\hbar\omega$ has its maximum value at $\hbar\omega_{max} = 1.59\, kT$.

With the energy per photon $\varepsilon_\gamma = \hbar\omega$, we find the energy of the photons per volume and per photon energy interval $d\hbar\omega$ in the solid angle interval $d\Omega$

$$\frac{de_\gamma(\hbar\omega)}{d\hbar\omega} = D_{\gamma,\Omega}(\hbar\omega) f_\gamma(\hbar\omega)\hbar\omega\, d\Omega = \frac{(\hbar\omega)^3\, d\Omega}{4\pi^3\hbar^3(c_0/n)^3} \frac{1}{\exp(\hbar\omega/kT) - 1} \tag{2.8}$$

The energy density per photon energy $de_\gamma/d\hbar\omega$ is called a spectrum. It has its maximum value at a photon energy of

$$\hbar\omega_{max} = 2.82\, kT \tag{2.9}$$

In the literature, another spectrum is often used, the energy density per wavelength $de_\gamma/d\lambda$ as a function of λ. With the relationship

$$\hbar\omega = h\nu = \frac{hc}{\lambda} \quad \text{from which follows} \quad d\hbar\omega = -\frac{hc}{\lambda^2}\, d\lambda \tag{2.10}$$

(Equation 2.8) is transformed into

$$\frac{de_\gamma(\lambda)}{d\lambda} = \frac{2hc\, d\Omega}{\lambda^5} \frac{1}{\exp[hc/(\lambda kT)] - 1} \tag{2.11}$$

We have left out the minus sign contained in Equation 2.10, which simply means that the photon energy $\hbar\omega$ decreases with increasing wavelength λ. The maximum value of $de_\gamma/d\lambda$ is at a wavelength

$$\lambda_{max} = \frac{hc}{4.965\,kT} = 0.2497\,\frac{\mu m\ eV}{kT}$$

To convert photon energies $\hbar\omega$ to wavelengths λ, we make use of Equation 2.10:

$$\hbar\omega\lambda = hc = 1.240\ eV\ \mu m \tag{2.12}$$

From Equation 2.8 the total energy of radiation per volume in the cavity, or the energy density, is

$$e_\gamma = \int_0^\infty \frac{(\hbar\omega)^3\, d\hbar\omega}{4\pi^3\hbar^3(c_0/n)^3[\exp(\hbar\omega/kT) - 1]} \int_0^{4\pi} d\Omega \tag{2.13}$$

Setting $x = \hbar\omega/kT$, the energy density is

$$e_\gamma = \frac{(kT)^4}{4\pi^3\hbar^3(c_0/n)^3} \underbrace{\int_0^\infty \frac{x^3\, dx}{e^x - 1}}_{\pi^4/15} 4\pi = \frac{\pi^2 k^4}{15\hbar^3(c_0/n)^3} T^4 \tag{2.14}$$

The same result is of course obtained by integrating Equation 2.11 over the wavelength.

With reference to Equation 2.7, the density of all photons in the cavity becomes

$$n_\gamma = \frac{(kT)^3}{4\pi^3\hbar^3(c_0/n)^3} \underbrace{\int_0^\infty \frac{x^2\, dx}{e^x - 1}}_{2.40411} 4\pi = \frac{2.40411\, k^3}{\pi^2\hbar^3(c_0/n)^3} T^3 \tag{2.15}$$

The average photon energy of blackbody radiation is thus given by

$$\langle\hbar\omega\rangle = \frac{e_\gamma}{n_\gamma} = 2.701\, kT \tag{2.16}$$

2.1.2
Energy Current Through an Area dA into the Solid Angle $d\Omega$

The energy density per solid angle inside the cavity is $e_\Omega = e_\gamma/4\pi$ for the entire spectrum. It is transported by photons, which move with the velocity c. We would like to find the energy current through a small hole in the cavity with surface area dA, directed into a small solid angle element $d\Omega$ normal to dA. Figure 2.3 shows that those Photons, which at time $t = 0$, are within the volume $dV = L^2\, d\Omega\, dL$ and are moving toward the hole will fly through the hole at time $t = L/c$ for a time interval $dt = dL/c$. With them the energy flows through the hole dA and into the solid angle element $d\Omega$. For an isotropic distribution, the fraction $d\Omega/4\pi = dA/4\pi L^2$ of the energy $dE = e_\gamma\, dV$ contained in the volume element dV flows through the hole during the time dt, and the energy current is

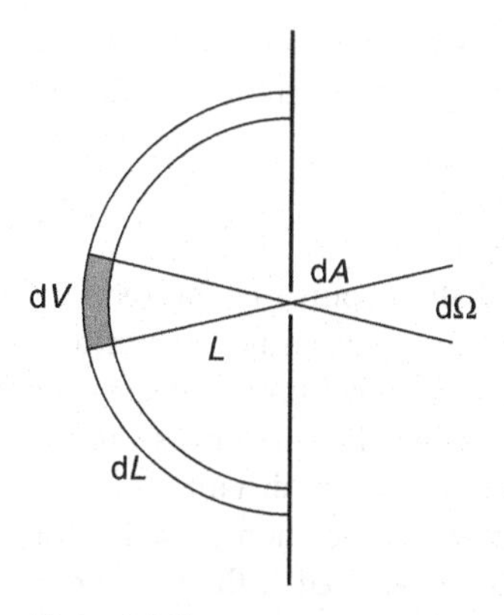

Figure 2.3 An energy current I_E originating from the volume element dV is flowing through the hole with area dA of the cavity, into the solid angle dΩ.

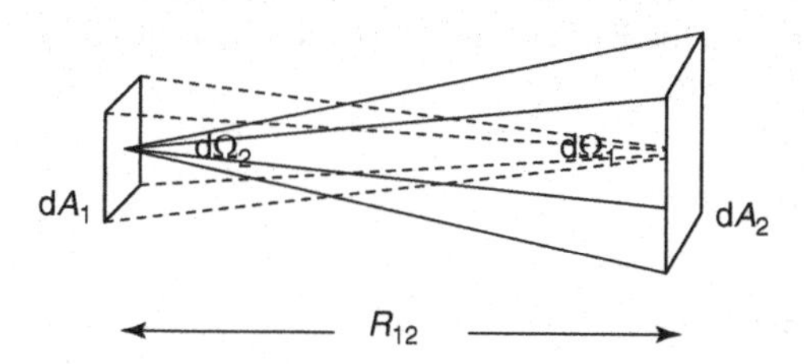

Figure 2.4 Solid angle dΩ_2 under which the receiver area dA_2 is seen from the emitter dA_1, and solid angle dΩ_1 under which dA_1 is seen from dA_2.

$$\mathrm{d}I_\mathrm{E} = \frac{e_\gamma}{4\pi}\frac{\mathrm{d}L}{\mathrm{d}t}\frac{\mathrm{d}A}{L^2}L^2\,\mathrm{d}\Omega = e_\Omega\, c\,\mathrm{d}A\,\mathrm{d}\Omega \tag{2.17}$$

and is as large as one would expect from the homogeneous energy density at the location of the hole. This equation also tells us that the point source radiator ($\mathrm{d}A = 0$), so popular in geometrical optics, would require an infinitely large energy density for a nonzero energy current and, therefore, cannot exist.

Let us now examine the propagation of the radiation toward a receiver. The energy current emitted from a radiating area $\mathrm{d}A_1$ toward a receiver area $\mathrm{d}A_2$ at a distance of R_{12} can be described from two different points of view (Figure 2.4):

- Viewed from the radiation source, the energy current $\mathrm{d}I_{E1}$ is emitted from all points of the surface area $\mathrm{d}A_1$ into the solid angle $\mathrm{d}\Omega_2 = \mathrm{d}A_2/R_{12}^2$.
- Viewed from the receiver, the energy current $\mathrm{d}I_{E2}$ is incident from the solid angle $\mathrm{d}\Omega_1 = \mathrm{d}A_1/R_{12}^2$ onto all points of the surface area $\mathrm{d}A_2$.

Both of course represent the same energy current

$$\mathrm{d}I_{E1} = e_{\Omega 1} c\,\mathrm{d}A_1\,\frac{\mathrm{d}A_2}{R_{12}^2} = \mathrm{d}I_{E2} = e_{\Omega 2} c\,\mathrm{d}A_2\,\frac{\mathrm{d}A_1}{R_{12}^2} \tag{2.18}$$

We see that the energy density per solid angle $e_{\Omega 1} = e_{\Omega 2}$ is the same at dA_1 and at dA_2 and thus does not change during the energy transport through the vacuum. In the same way, the energy current density per solid angle

$$j_{E,\Omega} = e_\Omega c$$

also does not change along the path from the radiation source to the receiver. The fact that the energy current dI_{E2} incident on the receiver surface decreases with increasing distance R_{12} from the emitter follows simply from the decreasing magnitude of the solid angle $d\Omega_1 = dA_1/R_{12}^2$ under which the emitter is seen.

Although we define solid angles by the size and distance of distant objects, it must be emphasized that the solid angle $d\Omega$ into which radiation is emitted at the location of the emitter (or from which radiation is received at the location of the receiver) is a local variable that describes the momentum distribution of the photons (with respect to direction) at this location.

According to Figure 2.5, the energy current from a surface element dA_s of the Sun incident upon a surface element dA_e of the Earth when dA_s and dA_e are both normal to a line connecting the Sun and the Earth over the distance R_{se} is given by

$$\begin{aligned} dI_{E,\text{Sun-Earth}} &= e_{\Omega,\text{Sun}} c \, d\Omega_e \, dA_s \\ &= e_{\Omega,\text{Sun}} c \frac{dA_e \, dA_s}{R_{se}^2} \\ &= e_{\Omega,\text{Sun}} c \, dA_e \, d\Omega_s \end{aligned} \tag{2.19}$$

It follows from the first expression that the energy current density per solid angle on the Sun at a point of the surface element dA_s is

$$j_{E,\Omega,\text{Sun}} = \frac{dI_{E,\text{Sun-Earth}}}{dA_s \, d\Omega_e} = e_{\Omega,\text{Sun}} c \tag{2.20}$$

In accordance with the third line of Equation 2.19 this is identical with the energy current density per solid angle on the Earth at a point on the surface element dA_e

$$j_{E,\Omega,\text{Earth}} = \frac{dI_{E,\text{Sun-Earth}}}{dA_e \, d\Omega_s} = e_{\Omega,\text{Sun}} c = j_{E,\Omega,\text{Sun}}$$

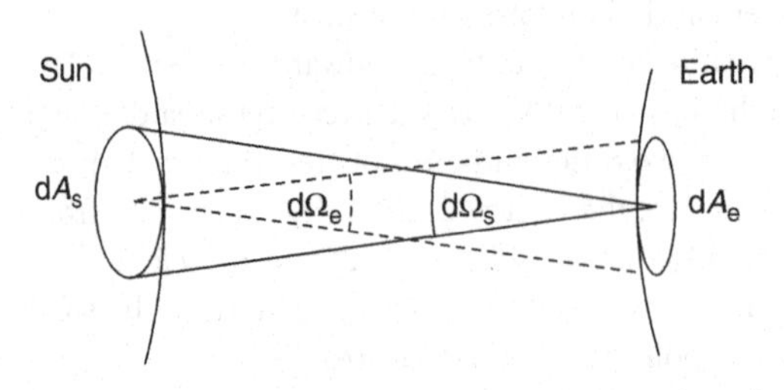

Figure 2.5 Solid angle $d\Omega_s$ under which the surface element dA_s on the Sun is viewed from the Earth, and solid angle $d\Omega_e$ under which the surface element dA_e on the Earth is viewed from the Sun.

This is an important result. Even though the energy current per surface area, the energy current density, decreases with distance from the Sun, it is of course much larger near the Sun than on the Earth; the energy current density per solid angle remains unchanged. Close to the Sun, the Sun simply subtends a larger solid angle than when viewed from a location on the Earth.

A long, thin tube, which when you look through it defines a small solid angle, can be used as an instrument to measure $j_{E,\Omega}$ as long as the emitter fills a larger solid angle than defined by the tube.

To find the energy current emitted from the entire Sun toward the Earth, we must integrate over the surface of the Sun, taking into account that not all surface elements are oriented normally to the line connecting the Earth and the Sun.

2.1.3
Radiation from a Spherical Surface into the Solid Angle dΩ

We can determine the energy current density dI_E that is radiated into the solid angle $d\Omega_e$ in a direction other than normal to the surface element dA_s by just looking at the Sun. In spite of its spherical shape, we see the Sun as a uniformly bright disc (apart from a slight drop in intensity very close to the edge). This enables us to conclude that each surface element on the Sun that *appears* to have the same size, when viewed from the Earth, in fact radiates the same energy current in our direction. The apparent size dA_s' of a surface element dA_s in Figure 2.6 is the projection of dA_s onto a plane normal to the line connecting the Earth and the Sun, that is,

$$dA_s' = dA_s \cos\vartheta$$

and the energy current emitted toward the Earth, (i.e. into $d\Omega_e$) is

$$dI_{E,e} = j_{E,\Omega}\, d\Omega_e\, dA_s \cos\vartheta$$

This dependence is called Lambert's law.

Figure 2.6 shows that all surface elements located on a ring around the line connecting the Earth and the Sun are positioned at the same angle in relation to this line. The radius of this ring is $r = R_s \sin\vartheta$, where R_s is the radius of the Sun. The area of the ring is

$$dA_s = 2\pi r R_s\, d\vartheta = 2\pi R_s^2 \sin\vartheta\, d\vartheta$$

Since only the hemisphere directed toward us radiates energy to the Earth, the total energy current that the Sun radiates to the Earth is

$$I_{E,e} = j_{E,\Omega}\, d\Omega_e \int_0^{\pi/2} 2\pi R_s^2 \sin\vartheta \cos\vartheta\, d\vartheta \tag{2.21}$$

Since $\cos\vartheta\, d\vartheta = d(\sin\vartheta)$ the integration is simple, and

$$I_{E,e} = j_{E,\Omega}\, d\Omega_e \pi R_s^2 \tag{2.22}$$

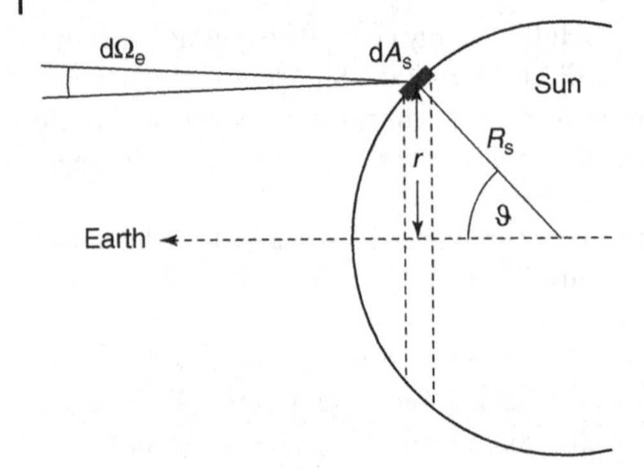

Figure 2.6 Energy current from the surface element dA_s on the surface of the Sun into the solid angle $d\Omega_e$ under which the Earth is viewed from the Sun.

This is exactly the result that we had expected: the energy current emitted from the Sun to the Earth is the same as that of a flat disc with a radius R_s oriented perpendicularly to the line connecting the Earth and the Sun.

2.1.4
Radiation from a Surface Element into a Hemisphere (Stefan–Boltzmann Radiation Law)

Let us now imagine a hemisphere of radius R surrounding the surface element dA from which radiation is emitted. All surface elements lying on a ring on the hemisphere are then seen from dA at the same angle relative to the normal to the surface element. The rest follows in the same way as in Section 2.1.3. The area of the ring $dO = 2\pi R^2 \sin\vartheta\, d\vartheta$ defines a solid angle element $d\Omega = dO/R^2 = 2\pi \sin\vartheta\, d\vartheta$. Integration over all energy currents $dI_E = j_{E,\Omega}\, d\Omega\, dA \cos\vartheta$ then yields the total energy current emitted from the surface element dA into the hemisphere

$$I_E = j_{E,\Omega}\pi\, dA \tag{2.23}$$

The energy current density emitted from the surface element dA into a hemisphere is

$$j_E = j_{E,\Omega}\pi = e_{\gamma,\Omega} c\pi = c\, e_\gamma/4 \tag{2.24}$$

Making use of Equation 2.14 for the energy density e_γ of black-body radiation then gives the Stefan–Boltzmann radiation law

$$j_E = \frac{\pi^2 k^4}{60\hbar^3 c^2} T^4 = \sigma T^4 \tag{2.25}$$

The value of the Stefan–Boltzmann constant $\sigma = 5.67 \times 10^{-8}\ \mathrm{W\,m^{-2}\,K^{-4}}$. The Stefan–Boltzmann law for the emission of black bodies was originally discovered experimentally in 1879 by Stefan and later derived theoretically by Boltzmann. We keep in mind that because of the dependence of the energy current density

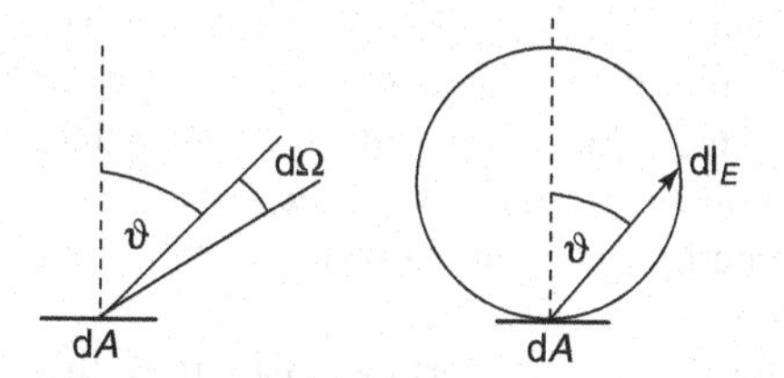

Figure 2.7 Dependence of the emitted energy current dI_E on the angle ϑ relative to the normal to the surface. The length of the arrow on the right is a measure of the magnitude of dI_E.

on $\cos\vartheta$ at the location of the emitter, as shown in Figure 2.7, the total energy current density emitted into the hemisphere ($\Omega = 2\pi$) is obtained by multiplying the energy current density per solid angle by π, as the effective solid angle for the emission of a planar black body into the hemisphere.

Equation 2.24 also allows us to write the energy density per solid angle in the form

$$e_\Omega = \frac{\sigma}{\pi c} T^4 \tag{2.26}$$

For a black body, the energy current density per solid angle is

$$j_{E,\Omega} = \frac{\sigma T^4}{\pi} \tag{2.27}$$

and the energy current density emitted normal to a surface into the small solid angle element $d\Omega$ is

$$j_E = \sigma T^4 \frac{d\Omega}{\pi} \tag{2.28}$$

As already mentioned, the dependence of the energy current on $\cos\vartheta$ means that a light-emitting surface appears equally bright at all angles.

As a consequence of this so-called Lambert's law for an emitting black, planar surface, the energy current density per solid angle observed from outside the surface $j_{E,\Omega}$ is independent of the angle ϑ at which the emitting surface is viewed. The dependence of the energy current density j_E on the angle of viewing results from a dependence of the solid angle subtended by the emitting area. This behavior precludes the recognition of any further details of a body by the emitted radiation, except for its contour.

This so-called Lambert behavior cannot be taken for granted. It applies strictly only for the surfaces of bodies that absorb all radiation incident upon them, i.e. black bodies. Weakly absorbing bodies such as a sheet of transparent plastic, in which low concentrations of dye molecules are dissolved, behave differently. These molecules emit light isotropically (e.g. as luminescence radiation when being excited by higher-energy photons). This light is not attenuated by absorption and remains isotropic. If we look onto the large surface of the pane, we see all molecules, just as if we look onto the edge of the pane, because as a result of their low concentration the dye molecules do not conceal each other. The eye perceives the same number of photons from each molecule, regardless of the direction from which we look at the pane. The edge therefore appears much brighter than the large surfaces. This property is common to luminescent plastic jewelry. Our interpretation is strictly valid only for materials with refractive index $n = 1$; for $n > 1$ the emission is less isotropic.

Very different designations are in use for energy transport by radiation. In this book, we will continue to use the terms:

Energy current	I_E	(measured in W)
Energy current density	j_E	(measured in W m^{-2})
Energy current density per solid angle	$j_{E,\Omega}$	(measured in W m^{-2}sr^{-1}).

These terms are physically clear and we think that the name for the transport of energy should not depend on the medium by which the energy is transported.

2.2
Kirchhoff's Law of Radiation for Nonblack Bodies

In the cavity of Figure 2.8 two plates are positioned opposite each other at a distance that is small compared to their lateral dimensions (different from the drawing). Plate 2 on the right is black, like the hole of a cavity, and absorbs all incident radiation completely ($a_2 = 1$). Plate 1 on the left represents a real material. It reflects (according to its reflectance r_1), transmits (according to its transmittance t_1), and absorbs (according to its absorptance a_1) the respective part of the radiation incident from the black plate $dj_{E,2}$ or from the black walls of the cavity $j_{E,w}$. Since reflection, transmission, and absorption depend on the photon energy, we restrict ourselves to the exchange of photons with an energy between $\hbar\omega$ and $\hbar\omega + d\hbar\omega$, by inserting a filter between the plates. Radiation with other photon energies is

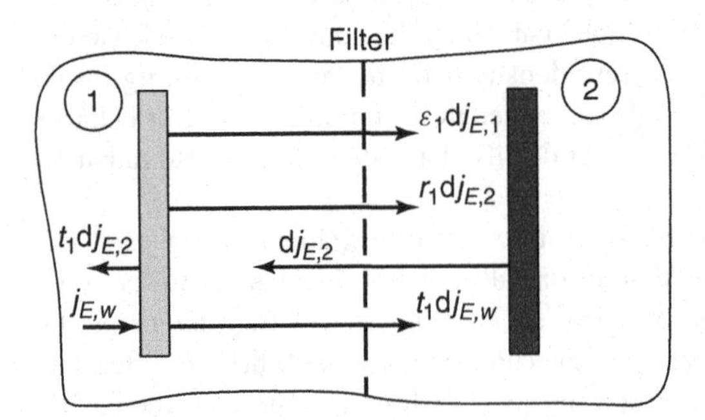

Figure 2.8 Exchange of radiation between two plates in a cavity with a filter in between, transmitting only photons with energy $\hbar\omega$.

perfectly reflected by the filter. The energy emitted by the black plate 2 in the energy interval $d\hbar\omega$ at the temperature T_2 with the energy current density $dj_{E,2}(\hbar\omega, T_2)$ follows from Planck's law of radiation in Equation 2.8 and is

$$dj_E(\hbar\omega, T_2) = \frac{d\Omega}{4\pi^3\hbar^3(c_0/n)^2} \frac{(\hbar\omega)^3}{\exp(\hbar\omega/kT_2) - 1} d\hbar\omega \tag{2.29}$$

The radiation incident on plate 1 is either reflected, transmitted, or absorbed, so that

$$r_1(\hbar\omega) + t_1(\hbar\omega) + a_1(\hbar\omega) = 1 \tag{2.30}$$

Plate 1 also emits radiation, which is $\varepsilon_1(\hbar\omega)\, dj_{E,1}$ and differs from the energy current $dj_{E,1}$ emitted by a black body of the same temperature T_1 by its (still unknown) emittance $\varepsilon_1(\hbar\omega)$.

In a steady state, each plate must emit the same amount of energy as it absorbs. We examine the energy balance for the black body, plate 2. The energy current incident onto the black body, plate 2, is composed of the energy reflected by plate 1, the energy that plate 1 emits at its temperature T_1, and the energy from the black cavity walls at the temperature T_w transmitted by plate 1. Since the absorptance of plate 2 is $a_2(\hbar\omega) = 1$, all the incident energy is absorbed and must be balanced by the emitted energy current $dj_{E,2}$, which is transmitted by the filter:

$$\begin{aligned} dj_{E,2}(\hbar\omega, T_2) = {} & r_1(\hbar\omega)\, dj_{E,2}(\hbar\omega, T_2) + \varepsilon_1(\hbar\omega)\, dj_{E,1}(\hbar\omega, T_1) \\ & + t_1(\hbar\omega)\, dj_{E,w}(\hbar\omega, T_w) \end{aligned} \tag{2.31}$$

Thermal equilibrium of course prevails in the cavity, and the exchange of radiation between the plates must not disturb this state. If the exchange of radiation were to cause differences in temperature, one could set up a thermoelectric generator for the generation of electrical energy, which would convert thermal energy completely into electricity. However, this would have to destroy entropy in contradiction with the second principle of thermodynamics.

With $T_1 = T_2 = T_W$ and consequently $dj_{E,1} = dj_{E,2} = dj_{E,w}$ in Equation 2.31, we obtain

$$r_1(\hbar\omega) + t_1(\hbar\omega) + \varepsilon_1(\hbar\omega) = 1 \tag{2.32}$$

We have left out the energy currents emitted by plate 2 toward the black cavity wall together with the energy current received by plate 2 from the wall, since they cancel each other. Comparing Equation 2.32 with Equation 2.30, we find

$$\varepsilon_1(\hbar\omega) = 1 - r_1(\hbar\omega) - t_1(\hbar\omega) = a_1(\hbar\omega) \tag{2.33}$$

The emissivity $\varepsilon(\hbar\omega)$ of a body at the photon energy $\hbar\omega$ is equal to its absorptance $a(\hbar\omega)$ at the same photon energy. This relationship is Kirchhoff's law of radiation stating that the energy current density emitted by a nonblack body in the energy interval $d\hbar\omega$ into the solid angle element $d\Omega$ is

$$\mathrm{d}j_E(\hbar\omega) = \frac{a(\hbar\omega)\,\mathrm{d}\Omega}{4\pi^3\hbar^3(c_0/n)^2} \frac{(\hbar\omega)^3}{\exp(\hbar\omega/kT) - 1}\,\mathrm{d}\hbar\omega \tag{2.34}$$

It is transported by the emitted photon current density

$$\mathrm{d}j_\gamma(\hbar\omega) = \frac{a(\hbar\omega)\,\mathrm{d}\Omega}{4\pi^3\hbar^3(c_0/n)^2} \frac{(\hbar\omega)^2}{\exp(\hbar\omega/kT) - 1}\,\mathrm{d}\hbar\omega \tag{2.35}$$

The absorptance $a(\hbar\omega)$ is characteristic of a body and is a function of the body's geometry. It describes the fraction of the incident light that is absorbed by the body. It is related to the absorption coefficient $\alpha(\hbar\omega)$, which is a material property and independent of the geometry of a body. Neglecting multiple reflections, the transmittance of a plate of thickness d is

$$t(\hbar\omega) = [1 - r(\hbar\omega)] \exp[-\alpha(\hbar\omega)d] \tag{2.36}$$

From $a = 1 - r - t$, we obtain the absorptance of a plate of thickness d

$$a(\hbar\omega) = [1 - r(\hbar\omega)]\{1 - \exp[-\alpha(\hbar\omega)d]\} \tag{2.37}$$

2.2.1 Absorption by Semiconductors

In the next chapter, we learn that solar cells are made from semiconductors. The absorption of solar radiation by solar cells is therefore characterized by a threshold photon energy $\hbar\omega_G$. Only photons with energies greater than $\hbar\omega_G$ can be absorbed. Photons with lower energies are either reflected or transmitted. Let us assume that reflection has been eliminated by an appropriate antireflection coating and that the semiconductor absorber is thick enough so that no photons with energy $>\hbar\omega_G$ are transmitted. The absorptance is then approximated by $a(\hbar\omega < \hbar\omega_G) = 0$ and $a(\hbar\omega \geq \hbar\omega_G) = 1$.

The energy current density absorbed by such a semiconductor plate is

$$j_{E,\text{abs}} = \frac{\mathrm{d}\Omega}{4\pi^3\hbar^3c^2} \int_{\hbar\omega_G}^{\infty} \frac{(\hbar\omega)^3}{\exp(\hbar\omega/kT) - 1}\,\mathrm{d}\hbar\omega \tag{2.38}$$

This integral can be evaluated. With $\hbar\omega/kT = x$ and $\hbar\omega_G/kT = x_G$ and rewriting

$$\frac{1}{\exp(x) - 1} = \frac{\exp(-x)}{1 - \exp(-x)} = \exp(-x)\sum_{i=0}^{\infty}\exp(-ix) = \sum_{i=1}^{\infty}\exp(-ix)$$

which is valid for $x > 0$, Equation 2.38 becomes

$$j_{E,\text{abs}} = \frac{\mathrm{d}\Omega\,(kT)^4}{4\pi^3\hbar^3c^2}\sum_{i=1}^{\infty}\int_{x_G}^{\infty} x^3\exp(-ix)\,\mathrm{d}x$$

Partial integration then gives

$$j_{E,\text{abs}} = \frac{\mathrm{d}\Omega\,(kT)^4}{4\pi^3\hbar^3c^2}\sum_{i=1}^{\infty}\exp(-ix_G)\left(\frac{x_G^3}{i} + \frac{3x_G^2}{i^2} + \frac{6x_G}{i^3} + \frac{6}{i^4}\right) \tag{2.39}$$

For $\hbar\omega_G \gg kT$, that is $x_G \gg 1$, only a few terms of this series are required to obtain sufficient accuracy. Accordingly, for the absorbed photon current density we obtain

$$j_{\gamma,\mathrm{abs}} = \frac{d\Omega (kT)^3}{4\pi^3\hbar^3c^2} \sum_{i=1}^{\infty} \exp(-ix_G) \left(\frac{x_G^2}{i} + \frac{2x_G}{i^2} + \frac{2}{i^3} \right) \tag{2.40}$$

2.3 The Solar Spectrum

Figure 2.9 shows the solar energy current density per photon energy interval as a function of the photon energy that is incident on the Earth before entering the atmosphere. It agrees well with a spectrum calculated from Equation 2.34 if we assume that the Sun is a black body and has a temperature T_S of 5800 K. At this temperature of the surface of the Sun, $kT_S = 0.5$ eV. The solar spectrum has a maximum at $\hbar\omega_{max} = 2.82\, kT_S = 1.41$ eV, that is, in the infrared part of the spectrum, as seen in Figure 2.9. This is just outside the visible range extending from 1.5 to 3 eV. The average energy of the solar photons is $2.7kT_S = 1.35$ eV.

Most often a different quantity is presented as the solar spectrum. It is the energy current density per wavelength interval as a function of the wavelength as shown in Figure 2.10.

This solar spectrum has a maximum at $\lambda_{max} = 0.5\ \mu m$, corresponding to a photon energy of $\hbar\omega = 2.48$ eV and distinctly different from the value $\hbar\omega_{max}$ found from the photon energy spectrum. Although both spectra are called solar spectra, they are completely different quantities and have different units. We will, therefore, use different names and call the energy current density per energy interval $dj_E/d\hbar\omega$

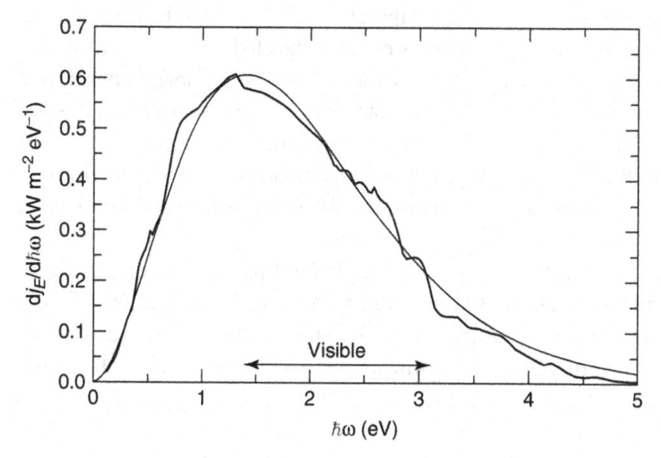

Figure 2.9 Energy current density per photon energy from the Sun as a function of the photon energy just outside the Earth's atmosphere (heavy line) compared with a black-body at a temperature of 5800 K (thin line).

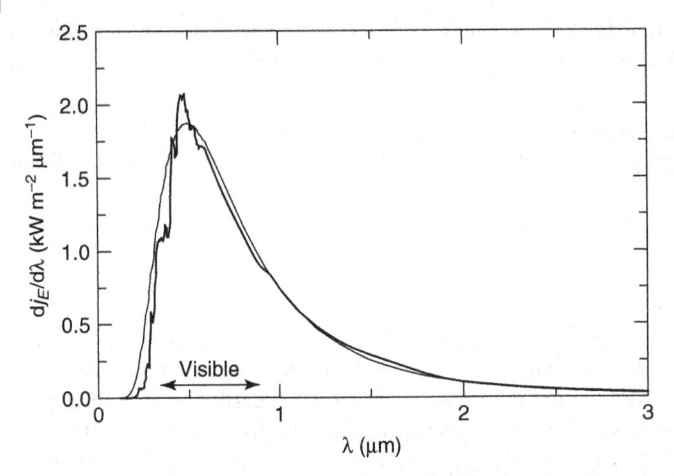

Figure 2.10 Energy current density per wavelength from the Sun as a function of the wavelength outside the Earth's atmosphere (heavy line) compared with a black body at a temperature of 5800 K (thin line).

as a function of the photon energy, shown in Figure 2.9, the energy spectrum, and the energy current density per wavelength interval as a function of the wavelength $dj_E/d\lambda$, shown in Figure 2.10, the wavelength spectrum.

The energy current density outside the atmosphere, that is, the integral over each of the curves in Figure 2.9 or 2.10, has a value of

$$j_{E,AM0} = 1353\ \mathrm{W\,m^{-2}} \tag{2.41}$$

The form of the spectrum in Figure 2.9 by itself is not sufficient to assert that the source, namely the Sun, actually has a temperature of $T_S = 5800$ K. Although the spectrum has its maximum at the photon energy expected for this temperature, it is also possible to produce an energy current density per photon energy with the form and magnitude as shown in Figure 2.9 with much colder light bulbs and appropriate filters. This is the basis for solar simulators used to test solar cells. The temperature of the radiation source, provided it is black, only follows from the spectrum after considering the solid angle $d\Omega$ subtended by the radiation source, which for the Sun is $\Omega_S = 6.8 \times 10^{-5}$.

Only the energy current density per solid angle (and per photon energy or per wavelength) is a measure of the temperature of a thermal emitter. For a solar simulator, radiating the same energy current density per photon energy onto an absorber, the radiation emerges from a much larger solid angle than for the Sun.

2.3.1
Air Mass

Solar radiation is partially absorbed during its passage through the atmosphere. The absorption is almost entirely caused by gases of low concentration in the infrared region of the solar spectrum, by water vapor (H_2O), carbon dioxide (CO_2), laughing gas (N_2O), methane (CH_4), fluorinated hydrocarbons, as well as by dust and, in the ultraviolet region of the spectrum, by ozone and oxygen. Absorption of course increases with the length of the path through the atmosphere and therefore with the mass of air through which the radiation passes. For a thickness l_0 of the atmosphere, the path length l through the atmosphere for radiation from the Sun incident at an angle α relative to the normal to the Earth's surface is given by

$$l = \frac{l_0}{\cos\alpha}$$

The ratio l/l_0 is called the *air-mass* coefficient. It characterizes the real solar spectrum resulting from the absorption of a layer of air of thickness l. The spectrum outside the atmosphere is designated by *AM*0 and that on the surface of the Earth for normal incidence by *AM*1. A typical spectrum for moderate climates is *AM*1.5, which corresponds to an angle of incidence of solar radiation of 48° relative to the surface normal.

The energy spectrum *AM*1.5 shown in Figure 2.11 is regarded as the standard spectrum for measuring the efficiency of solar cells used terrestrially, i.e. on the surface of the Earth. The integral over this spectrum, the energy current density onto a surface normal to the Sun for a cloudless sky, is defined to be

$$j_{E,AM1.5} = 1.0\,\mathrm{kW\,m^{-2}}$$

This maximum energy current density is only negligibly larger for the *AM*1.0 spectrum. The maximum energy current density varies only slightly from the tropics to the moderate zones. The differences in the amount of energy incident onto a horizontal surface in one year are, however, much greater. In Germany

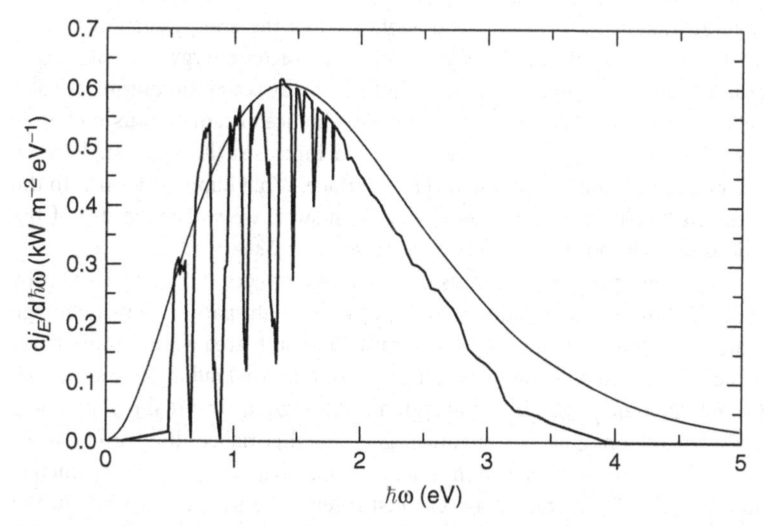

Figure 2.11 The *AM*1.5 spectrum (heavy line) compared with a black body at a temperature of 5800 K (thin line).

this is around 1000 kWh $m^{-2}a^{-1}$. We therefore speak of 1000 Sun-hours (with 1 kW m^{-2}) per year.

Averaged over the year, the mean energy current density in Germany is 115 W m^{-2}, and roughly a factor of 10 less than the maximum energy current density of the *AM*1.5 spectrum. The greatest annual amount of incident solar energy is found in Saudi Arabia, with a value of around 2500 kWh $m^{-2}a^{-1}$. This corresponds to an average energy current density of 285 W m^{-2}. The average value over the entire Earth is 230 W m^{-2}.

For the *AM*0 spectrum, outside the atmosphere, the energy current density averaged over the entire Earth is easy to calculate. We only have to divide the energy current $I_E = 1353 \times \pi R^2_{Earth}$ W m^{-2} incident upon the Earth by the total surface area of the Earth ($4\pi R^2_{Earth}$) to obtain the value $\langle j_e \rangle = 1/4 \times 1353$ W $m^{-2} = 338$ W m^{-2} for the average energy current density, if there were no atmosphere.

2.4 Concentration of the Solar Radiation

Viewed from the Earth, the Sun has an angular diameter α_S of 32′, corresponding to a solid angle of

$$\Omega_S = 2\pi \int_0^{\alpha_S/2} \sin\vartheta \, d\vartheta = 2\pi \left(1 - \cos\frac{\alpha_S}{2}\right) = 6.8 \times 10^{-5} \qquad (2.42)$$

It is due to this small value of the solid angle Ω_S that the energy current density on the Earth has a value of only 1 kW m^{-2}. When a greater energy current density is needed in order to obtain higher temperatures or increase the output of solar cells, it is necessary to focus the nearly parallel incidence of radiation using lenses or mirrors. This reduces the irradiated area and increases the solid angle from which the radiation impinges on the receiver surface. What happens then with the energy current density per solid angle? For an answer, we will make use of the deliberations of Helmholtz and Clausius from the year 1864.

Figure 2.12 shows two area elements dA_1 and dA_2, projected onto each other by an imaging system. Because of the reversibility of the light path, dA_2 is the image of dA_1 and dA_1 is the image of dA_2. We assume that both area elements are black bodies which, without any further illumination, emit thermal radiation toward each other through the transparent imaging system. Although the thermal radiation will be redirected by the imaging system, once again, the thermal equilibrium between dA_1 and dA_2 must not be disturbed in order to avoid violating the second principle of thermodynamics. Thus, the area dA_2 must receive exactly the same amount of radiation from the lens or the imaging system as it emits toward it. This equality of absorbed and emitted energy currents must exist not only for any photon energy,

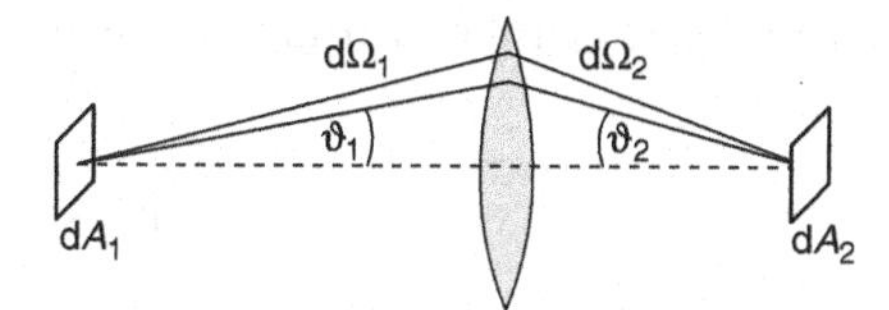

Figure 2.12 An imaging system projecting two area elements onto each other by redirecting their thermally emitted radiation must not disturb the thermal equilibrium between them.

as in Section 2.2, but also for any direction, that is, any solid angle element, as defined by an area element on the surface of the lens. By redirecting the radiation from dA_1, the imaging system (the lens) must therefore behave like a black body, emitting radiation toward dA_2 with the temperature of dA_1. But in contrast to a real black body, it selectively radiates only in the direction of dA_2 and in no other direction. We have already seen that for the propagation of radiation in free space the energy current density per solid angle remains unchanged and is therefore exactly the same at the location of the lens in the direction toward dA_2 as at the location of dA_2 from the direction of the lens. Owing to thermal equilibrium, this is of course exactly the same as the energy current density per solid angle at the location of dA_1. *During passage through a nonabsorbing and nonemitting ideal imaging system, the energy current density per solid angle remains unchanged.*

The reversibility of the light path also implies thermodynamic reversibility. This in turn means that no entropy is created during passage through an imaging system. And in fact the conservation of the energy current density per solid angle in regions with the index of refraction n, with simultaneous conservation of the energy per photon, is identical with the conservation of entropy. More precisely, for the propagation of radiation not attenuated by absorption or scattering, even when the index of refraction n changes, the occupation probability of the photon states f_γ in Equation 2.3 remains the same.

2.4.1
The Abbé Sine Condition

The energy current emitted in Figure 2.12 from dA_1 in the direction (ϑ_1, φ_1) into the solid angle element $d\Omega_1 = \sin\vartheta_1 \, d\vartheta_1 \, d\varphi_1$ is

$$dI_{E,1} = j_{E,\Omega,1} \, dA_1 \cos\vartheta_1 \sin\vartheta_1 \, d\vartheta_1 \, d\varphi_1 \tag{2.43}$$

This energy current is incident upon dA_2 and as a result of thermal equilibrium is identical with the energy current emitted from dA_2 in the direction (ϑ_2, φ_2) into the solid angle element $d\Omega_2 = \sin\vartheta_2 \, d\vartheta_2 \, d\varphi_2$

$$dI_{E,2} = j_{E,\Omega,2} \, dA_2 \cos\vartheta_2 \sin\vartheta_2 \, d\vartheta_2 \, d\varphi_2 = dI_{E,1} \tag{2.44}$$

To generalize the condition in Equation 2.44 by allowing for different media with different indices of refraction in front of and behind the lens in regions 1 and 2, we have to take into account Equation 2.34 for the energy current densities per solid angle,

$$\frac{j_{E,\Omega,1}}{j_{E,\Omega,2}} = \frac{c_2^2}{c_1^2} = \frac{n_1^2}{n_2^2} \tag{2.45}$$

The entire energy current absorbed by the area element dA_2 or emitted by it toward the lens is obtained by integrating over the entire lens or over the angle ϑ_2 from 0 to v, where v is the angle relative to the optical axis at which the edge of the lens is viewed from dA_2, and over φ_2 from 0 to 2π

$$I_{E,2} = j_{E,\Omega,2}\, dA_2 \pi \sin^2 v \tag{2.46}$$

In exactly the same way, we find the entire energy current emitted by the area element dA_1 toward the lens or obtained from the lens to be

$$I_{E,1} = j_{E,\Omega,1}\, dA_1 \pi \sin^2 u \tag{2.47}$$

where u is the angle at which the edge of the lens is viewed from dA_1.

Since the two energy currents must be equal due to thermal equilibrium, and using Equation 2.45, we obtain the Abbé sine condition for optical imagery

$$n_1^2\, dA_1 \sin^2 u = n_2^2\, dA_2 \sin^2 v \tag{2.48}$$

The basis of this condition is the conservation of the energy current density per solid angle in regions with the same index of refraction n, even when it is passing through imaging systems. This law of conservation is in no way trivial. As it turns out, it contradicts the simple imaging laws of geometrical optics.

2.4.2 Geometrical Optics

In accordance with geometrical optics, for the imaging of the Sun by a lens with radius r_L producing an image with area A_B as in Figure 2.13, we expect that the energy current incident upon the lens is also incident upon the image A_B of the Sun. The energy current incident upon the lens is

$$I_E = j_{E,\text{Sun}} \pi r_L^2$$

where $j_{E,\text{Sun}}$ is the energy current density of the solar radiation at the location of the lens. The energy current density in the image A_B is then

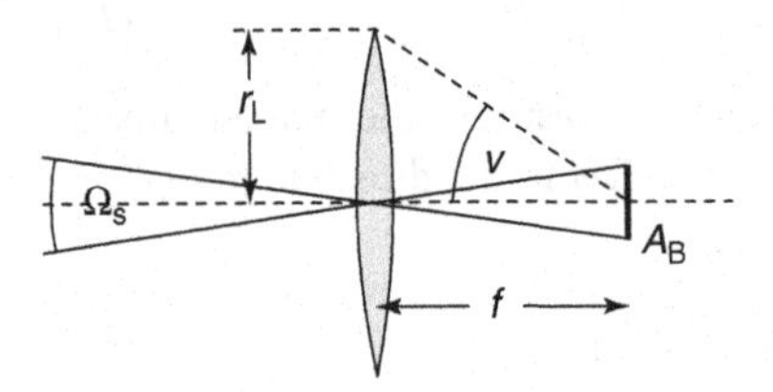

Figure 2.13 Image of the Sun with an area A_B in the focal plane of a lens with radius r_L and focal length f.

$$j_{E,\text{image}} = \frac{I_E}{A_B} = \frac{j_{E,\text{Sun}}\pi r_L^2}{A_B}$$

The concentration factor C is a measure of the enhancement of the energy current density by the lens and, according to geometrical optics, is therefore

$$C = \frac{j_{E,\text{image}}}{j_{E,\text{Sun}}} = \frac{\pi r_L^2}{A_B}$$

The size of the image A_B, formed in the focal plane because of the large distance from the Sun, depends only on the focal length f of the lens and not on its size. Since the aperture angle Ω_S of a bundle of rays incident upon the middle of the lens remains unchanged during passage through the lens, we have on the image side $\Omega_S = A_B/f^2$.

Expressing the radius of the lens as

$$r_L = f \tan \nu$$

in terms of the maximum angle ν of incidence onto the image, we finally arrive at

$$C = \frac{\pi}{\Omega_S} \tan^2 \nu \tag{2.49}$$

In agreement with the assumption that the energy current incident upon the lens is also incident upon the image of the Sun, the concentration factor becomes arbitrarily large when the lens is made arbitrarily large. As an example, let us imagine a lens with $\nu > 45°$, that is $r_L > f$. For this lens $C > \pi/\Omega_S$.

The ratio $2r_L/f$ represents the f-number of lenses, expressed as the ratio $1:(f/2r_L)$. $1:1.4$ or $1:2.0$ or $1:2.8$ are typical values for fast camera lenses. In the example above, the f-number would be $>1:0.5$. Certainly, no one has ever heard of such a lens, because it would in fact violate the second law of thermodynamics. For this lens, according to Equation 2.28 and Equation 2.49 the energy current density in the image of the Sun would be

$$j_{E,\text{image}} = C j_{E,\text{Sun}} > \frac{\pi}{\Omega_S}\frac{\Omega_S}{\pi}\sigma T_S^4$$

so that

$$j_{E,\text{image}} > \sigma T_S^4$$

and therefore greater than on the surface of the Sun where it is σT_S^4.

Under steady state conditions, a black body in the image of the Sun, with a perfectly reflecting rear surface and therefore unable to emit radiation from this side, would have to emit from the front side the absorbed energy current density $j_{E,\text{image}} = \sigma T_{\text{image}}^4 > \sigma T_S^4$. Its temperature T_{image} would then have to be greater than that of the Sun T_S. The emitted energy current $I_{E,\text{image}}$ also implies the emission of an entropy current. According to Boltzmann, the entropy current emitted by a black body of temperature T_{image} is related to the emitted energy current by

$$I_{S,\text{image}} = \frac{4}{3}\frac{I_{E,\text{image}}}{T_{\text{image}}}$$

This would be less than the entropy current absorbed together with the same energy current $I_{E,\text{image}}$

$$I_{S,\text{abs}} = \frac{4}{3}\frac{I_{E,\text{image}}}{T_{\text{Sun}}}$$

This would be possible only if entropy were continuously destroyed in the image of the Sun. However, the second law of thermodynamics forbids this.

2.4.3 Concentration of Radiation Using the Sine Condition

The correct determination of the enhancement of the energy current density by the use of concentrating optics takes into account the conservation of the energy current density per solid angle $j_{E,\Omega,\text{Sun}}$. Accordingly, viewed from the image of the Sun the entire lens appears to be as bright as the Sun (do not try this out!) and fills a cone with the angle ν. According to Equation 2.46, the energy current density in the image of the Sun is

$$j_{E,\text{image}} = j_{E,\Omega,\text{Sun}}\pi \sin^2 \nu$$

With $j_{E,\text{Sun}} = \Omega_S j_{E,\Omega,\text{Sun}}$ the concentration factor becomes

$$C = \frac{j_{E,\text{image}}}{j_{E,\text{Sun}}} = \frac{\pi}{\Omega_S}\sin^2 \nu \tag{2.50}$$

The maximum concentration is found for $\nu = 90^\circ$

$$C_{\max} = \frac{\pi}{\Omega_S} = 46\,200 \tag{2.51}$$

The contradiction between Equation 2.49, based on geometrical optics of an undistorted image, and Equation 2.50, based on thermodynamics and the impossibility of destroying entropy, can be resolved for imaging systems with curved principal "planes". The maximum possible f-number is 1:0.5 and requires a principal image-side plane in the form of a hemisphere, curved around the focal point. This is, however, not totally achievable.

The maximum concentration can nevertheless be obtained by other means. It ensures that an absorbing body reaches the temperature of the Sun, if it loses energy only by emission toward the Sun. In this situation it is in radiative equilibrium with the Sun. Let us assume that the absorber is already at the temperature of the Sun. It would then remain at this temperature if it were within a cavity with walls at the temperature of the Sun, that is, as though it could only see the Sun. It would also remain at the temperature of the Sun if it were located in a cavity with perfectly reflecting walls, that is, as though it could only see itself. Figure 2.14 shows such an arrangement in which a body is in radiative equilibrium with the Sun. The lens placed in the wall of a reflecting cavity produces an image of the Sun that is at least as large as the cross-section of the absorber. Since the absorber then sees only the Sun or itself, it would reach the temperature of the Sun if there were perfect mirrors. According to this consideration, the structural principle of concentrators can now be defined very simply. A good concentrator must be designed so as to direct toward the Sun the greatest possible amount of thermal radiation emitted by the absorber.

The ideal arrangement shown in Figure 2.14 permits the determination of the maximum efficiency with which solar energy can be converted into electrical energy.

2.5 Maximum Efficiency of Solar Energy Conversion

To use the incident solar energy in the arrangement of Figure 2.14, heat must be extracted from the absorber and conducted to a heat engine. Using an ideal heat engine, i.e. a Carnot engine, we obtain the greatest possible amount of electrical energy. Owing to the extraction of energy, the temperature of the absorber T_A will be lower than that of the Sun T_S. It must be lower so that it does not emit as much energy as it absorbs.

The useful net absorbed energy current $I_{E,\text{util}}$ that can be extracted from the absorber of cross-sectional area A_A is the difference between the absorbed energy current $I_{E,\text{abs}}$ and the energy current $I_{E,\text{emit}}$ emitted through the lens toward the Sun. Ideally, both fill the same solid angle Ω_L, at which the lens is viewed from the absorber. For a more general discussion, we will allow for different solid angles, Ω_{abs} for the incident radiation and Ω_{emit} for the emitted radiation:

$$I_{E,\text{util}} = I_{E,\text{abs}} - I_{E,\text{emit}} = \sigma\left(\frac{\Omega_{\text{abs}}}{\pi}T_S^4 - \frac{\Omega_{\text{emit}}}{\pi}T_A^4\right)A_A$$

We can now define the efficiency for obtaining $I_{E,\text{util}}$ as

$$\eta_{\text{abs}} = \frac{I_{E,\text{util}}}{I_{E,\text{abs}}} = 1 - \frac{I_{E,\text{emit}}}{I_{E,\text{abs}}} = 1 - \frac{\Omega_{\text{emit}}}{\Omega_{\text{abs}}}\frac{T_A^4}{T_S^4}$$

This relation states that for a high efficiency of net absorption η_{abs} the temperature T_A of the absorber should be as low as possible to avoid the emission of radiation.

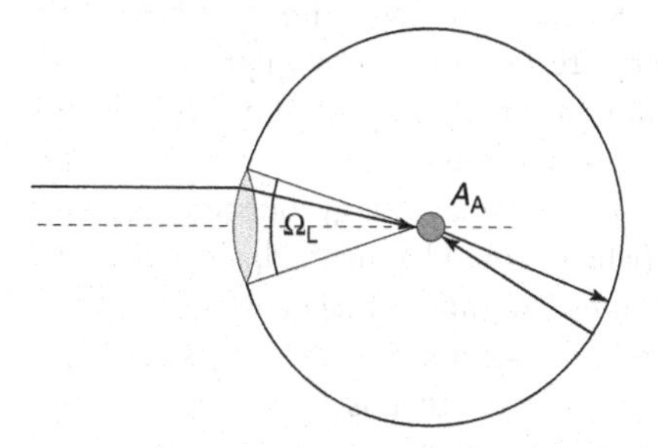

Figure 2.14 A lens projects an image of the Sun onto an absorber in a perfectly reflecting cavity.

Along with the heat energy current $I_{E,\text{util}}$, the entropy current $I_S = I_{E,\text{util}}/T_A$ is transferred from the absorber to the heat engine. According to the second law of thermodynamics, entropy cannot be destroyed, but it can in fact be created. In an ideal Carnot engine, entropy is conserved. Since the absorbed entropy current I_S cannot be stored in the engine, it must be transported to another container, a heat reservoir. If this accepts the entropy at the temperature T_0, the entropy current will be accompanied by a heat energy current $T_0 I_S$ from the engine to this heat reservoir. Only the difference between the absorbed energy current and that given to the reservoir is supplied by the Carnot engine as an entropy-free energy current, e.g. in the form of electrical energy $I_{E,\text{el}}$,

$$I_{E,\text{el}} = I_{E,\text{util}} - T_0 I_S = T_A I_S - T_0 I_S$$

The efficiency for this process is

$$\eta_C = \frac{I_{E,\text{el}}}{I_{E,\text{util}}} = \frac{T_A I_S - T_0 I_S}{T_A I_S} = 1 - \frac{T_0}{T_A} \tag{2.52}$$

The efficiency of such an ideal heat engine is known after its discoverer as the Carnot efficiency. For a high Carnot efficiency, the temperature T_A of the absorber should be as high as possible.

The overall efficiency η_{bC} for the conversion of solar energy to entropy-free energy, e.g. electrical energy, by a black-body absorber combined with a Carnot engine is

$$\eta_{bC} = \frac{I_{E,\text{el}}}{I_{E,\text{abs}}} = \frac{I_{E,\text{el}}}{I_{E,\text{util}}} \frac{I_{E,\text{util}}}{I_{E,\text{abs}}} = \eta_{\text{abs}} \eta_C$$

$$\eta_{bC} = \left(1 - \frac{\Omega_{\text{emit}}}{\Omega_{\text{abs}}} \frac{T_A^4}{T_S^4}\right) \left(1 - \frac{T_0}{T_A}\right) \tag{2.53}$$

Figure 2.15 displays η_{bC} as a function of T_A for $T_S = 5800$ K and $T_0 = 300$ K for maximum concentration, i.e. equal solid angles of incident and emitted radiation as seen in Figure 2.14. The overall efficiency has a maximum value of 0.85 for an absorber temperature $T_A = 2478$ K. This high efficiency value demonstrates that solar energy is very high-quality energy because of the high temperature of the Sun.

Since the thermal radiation from the Sun at a temperature $T_S = 5800$ K is absorbed by the absorber at a lower temperature T_A, the absorption process creates entropy. We will accept this, because we want to obtain the greatest possible energy current, and the heat radiated back to the Sun is lost to us.

An interesting aspect arises if we had to pay for solar energy, but could also get a refund for energy returned to the Sun (which would be justified, since it would actually prolong the Sun's lifetime). Under these conditions maximal concentration must be chosen along with a temperature of the absorber that is only slightly (by $\mathrm{d}T$) smaller than T_S. The net absorbed energy current then is

$$I_{E,\text{abs}} = \left[\frac{\Omega_{\text{abs}}}{\pi} \sigma T_S^4 - \frac{\Omega_{\text{emit}}}{\pi} \sigma \left(T_S - \mathrm{d}T\right)^4\right] A_A \approx 4 \frac{\Omega_{\text{abs}}}{\pi} \sigma T_S^3 \, \mathrm{d}T A_A$$

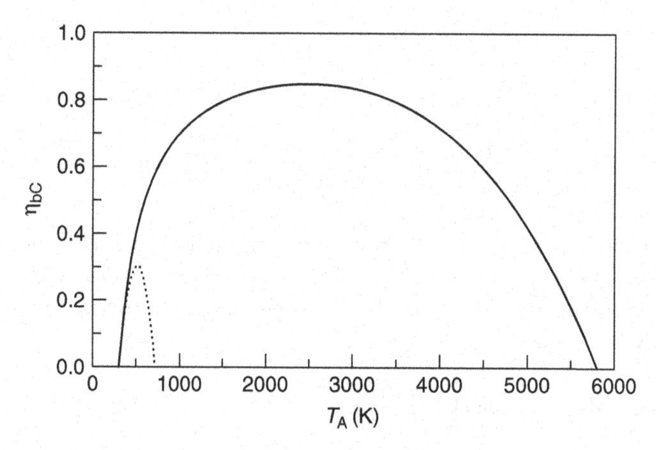

Figure 2.15 Efficiency η_{bC} for the conversion of solar heat energy by a black absorber and a Carnot engine as a function of the absorber temperature T_A, for maximum concentration ($\Omega_{emit} = \Omega_{abs}$, heavy line) and for nonconcentrated radiation ($\Omega_{emit} = \pi, \Omega_{abs} = 6.8 \times 10^{-5}$, broken line).

Since this is the energy current for which we would have to pay, the absorption efficiency is $\eta_{abs} = 1$. The net absorbed entropy current is

$$I_{S,abs} = \frac{4}{3}\frac{\Omega_{abs}}{\pi}\sigma\left[T_S^3 - (T_S - dT)^3\right]A_A \approx 4\frac{\Omega_{abs}}{\pi}\sigma T_S^2\, dT A_A$$

This time, the absorption process does not generate entropy, since the ratio of absorbed energy and entropy currents is equal to the Sun's temperature T_S. It is not surprising that the overall efficiency for this reversible process, including the Carnot engine, is $\eta = 1 - T_0/T_S$. The output of the engine is, however, rather small, since almost all of the incident solar radiation is emitted back to the Sun and an efficiency that relates the energy output not to the net absorbed but to the incident energy current is almost zero.

A slight improvement in efficiency over the combination of a black absorber and a Carnot engine is found for a combination of a monochromatic absorber and a Carnot engine. However, very (infinitely) many absorbers, each for a different photon energy interval, are needed to cover the whole solar energy spectrum. Each absorber would have its own Carnot engine and operate at its own optimal temperature.

We start with a single monochromatic absorber, absorbing all incident solar photons that are in a narrow energy interval $d\hbar\omega$ around a photon energy of, e.g. $\hbar\omega = 1.5$ eV and emitting photons in the same energy interval according to its temperature. Again, we assume maximal concentration of the incident solar radiation, i.e. equal solid angles for incident and emitted radiation. Using Planck's equation for the photon current density per photon energy in Equation 2.35, the net absorbed photon current, the difference between absorbed and emitted photon currents is

$$dj_{\gamma,\text{net}}(\hbar\omega) = dj_{\gamma,\text{abs}} - dj_{\gamma,\text{emit}}$$
$$= m\left[\frac{(\hbar\omega)^2}{\exp(\hbar\omega/kT_S) - 1} - \frac{(\hbar\omega)^2}{\exp(\hbar\omega/kT_A) - 1}\right] d\hbar\omega \qquad (2.54)$$

where the factor m contains all the constants including the solid angle. Each net absorbed photon carries its energy $\hbar\omega$ to the Carnot engine, which produces from it entropy-free energy of an amount of $\hbar\omega(1 - T_0/T_A)$ per photon, resulting in an energy current delivered by the Carnot engine of

$$dj_{E,\text{Carnot}}(\hbar\omega) = dj_{\gamma,\text{net}}(\hbar\omega)\,\hbar\omega\left(1 - \frac{T_0}{T_A}\right)$$

Both the net absorbed photon current and the free energy per photon depend on the absorber temperature T_A. By varying T_A we find pairs of absorbed photon current and free energy per photon, which are plotted in Figure 2.16 as the thick line. The current of free energy delivered by the Carnot engine is the area of a rectangle formed by the x- and y-values of each point on this curve. The hatched rectangle has the largest possible area and belongs to the point of maximum power on the curve. This point of maximum power follows from the maximum of the energy current $dj_{E,\text{Carnot}}(\hbar\omega)$ shown by the thin line.

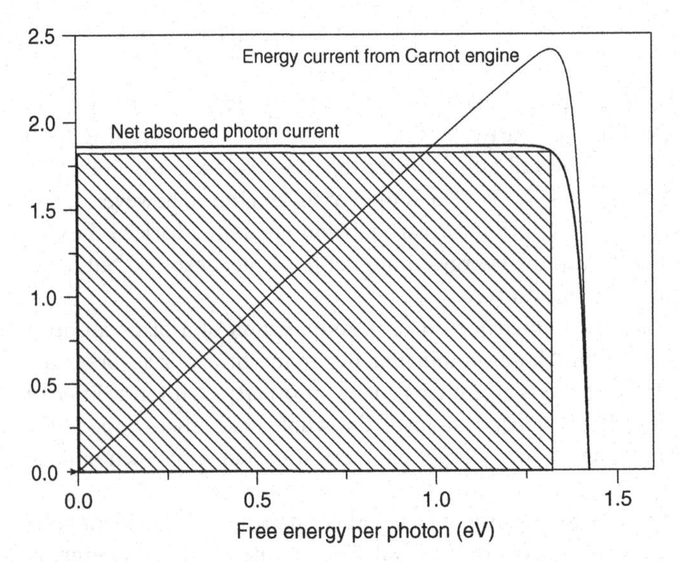

Figure 2.16 Net absorbed photon current of 1.5 eV photons as a function of the free energy per photon produced by a Carnot engine. The hatched rectangle, the largest rectangle for any point on the photon current curve, belongs to the point of maximum power, which also follows from the maximum of the energy current curve.

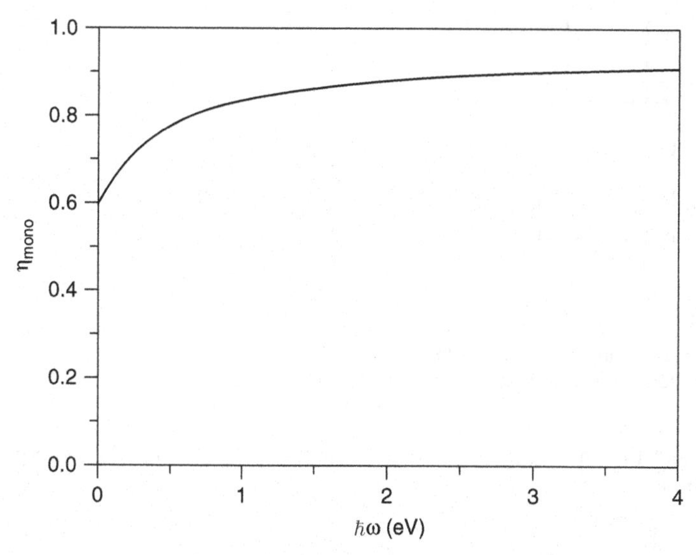

Figure 2.17 Efficiency of monochromatically absorbing Carnot engines as a function of the photon energy for fully concentrated black-body radiation of 5800 K.

The efficiency of energy conversion is found by dividing the energy current $dj_{E,\text{Carnot}}(\hbar\omega)$ by the incident monochromatic energy current $\hbar\omega\, dj_{\gamma,\text{abs}}(\hbar\omega)$. This efficiency is shown as a function of the photon energy in Figure 2.17 for maximum concentration. For the conversion of the whole solar spectrum, many monochromatic conversion engines have to operate in parallel. The entropy-free energy obtained from them is the integral over the energy current per photon energy interval weighted with the appropriate monochromatic efficiency. For the overall efficiency, this output energy current must be divided by the incident energy current, resulting in a value of 86%, which is slightly higher than the 85% obtained with a black absorber and a single Carnot engine.

So far, efficient energy conversion has had to avoid the emission of photons toward the Sun as far as possible. The temperature of the absorber, therefore, had to be substantially lower than the temperature of the Sun. This lower temperature of the absorber, however, gives rise to unwanted entropy generation, which could only be avoided when the absorber has the same temperature as the Sun, where the net absorbed energy current is zero. In this dilemma, we might ask ourselves how high the efficiency would be if it were possible for a black body at a temperature $T_A < T_S$ to absorb the radiation from the Sun without creating entropy. This efficiency is called the *Landsberg efficiency*. It follows from a balance of absorbed and emitted energy and entropy currents under the condition of reversibility as indicated in Figure 2.18.

The entropy absorbed with solar radiation is given off in two ways. One part is emitted back to the Sun, together with the thermal radiation at the temperature T_A, and the remaining part goes to a heat reservoir at the ambient temperature T_0, implying a loss of energy.

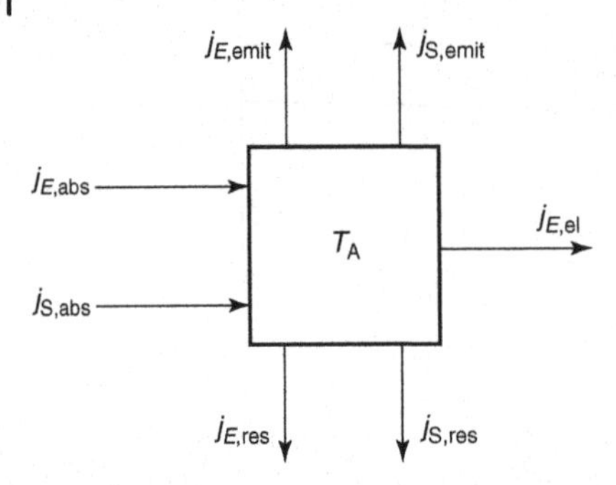

Figure 2.18 Balance of absorbed and emitted energy and entropy currents for a reversible heat engine working with radiation.

Assuming that the absorbing engine sees only the Sun, the absorbed energy current density is

$$j_{E,\,abs} = \sigma T_S^4$$

For black-body radiation, the density of the entropy current absorbed is

$$j_{S,\,abs} = \frac{4}{3}\frac{j_{E,\,abs}}{T_S} = \frac{4}{3}\sigma T_S^3$$

The energy current density emitted by the absorber with temperature T_A (to the Sun) is given by

$$j_{E,\,emit} = \sigma T_A^4$$

and the entropy current density emitted is

$$j_{S,\,emit} = \frac{4}{3}\sigma T_A^3$$

In order to get rid of all the entropy absorbed, the entropy current density $j_{S,\,res}$ is transferred to the reservoir. With the entropy the energy current density

$$j_{E,\,res} = T_0 j_{S,\,res}$$

is therefore transported to the reservoir and lost for conversion. The condition of reversibility, that is, the conservation of entropy, takes the form

$$j_{S,\,abs} = j_{S,\,emit} + j_{S,\,res}$$

This leads to the relationships

$$j_{S,\,res} = \frac{4}{3}\sigma\left(T_S^3 - T_A^3\right)$$

and

$$j_{E,\,res} = \frac{4}{3}\sigma T_0\left(T_S^3 - T_A^3\right)$$

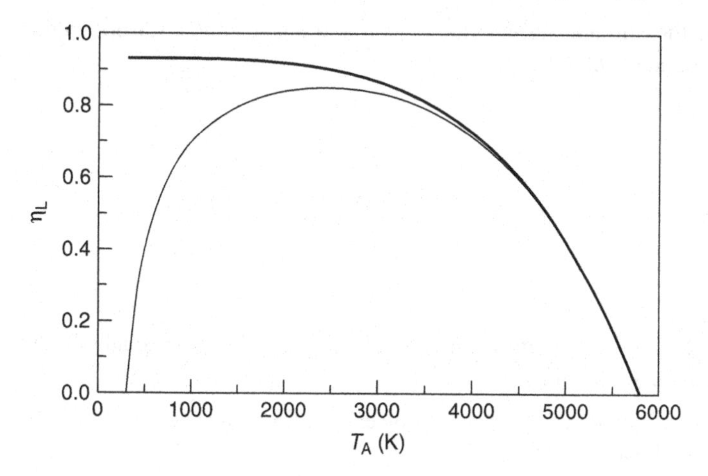

Figure 2.19 Landsberg efficiency η_L (heavy line) as a function of the absorber temperature T_A compared with the efficiency η_{bC} (thin line) of a solar heat engine with a single black absorber.

The entropy-free, utilizable energy current density $j_{E,\text{el}}$ is

$$j_{E,\text{el}} = j_{E,\text{abs}} - j_{E,\text{emit}} - j_{E,\text{res}}$$

The Landsberg efficiency is therefore

$$\begin{aligned}\eta_L &= \frac{j_{E,\text{el}}}{j_{E,\text{abs}}} = 1 - \frac{\sigma T_A^4 + 4/3\sigma T_0(T_S^3 - T_A^3)}{\sigma T_S^4} \\ &= 1 - \frac{T_A^4}{T_S^4} - \frac{4}{3}\frac{T_0}{T_S}\left(1 - \frac{T_A^3}{T_S^3}\right) \qquad (2.55)\end{aligned}$$

Figure 2.19 compares the Landsberg efficiency η_L with the efficiency η_{bC} of a single black absorber combined with a Carnot engine. The lower the absorber temperature T_A, the greater is η_L. However, it is highly questionable whether the reversible absorption of radiation of a temperature T_S by a black body of a temperature $T_A < T_S$ is permitted by nature. In this respect, the Landsberg efficiency sets an upper limit to any kind of solar energy conversion.

Later, when looking at semiconductors, we will discover absorbers capable of absorbing the incident radiation without creating entropy, even though their temperature is much lower than that of the radiation. Nevertheless, in these states they emit photons with a chemical potential greater than zero. This is different from thermal radiation, for which photons have a chemical potential equal to zero. In spite of this, we cannot take advantage of this property of semiconductors because, even though their temperature is lower, in order to avoid the creation of entropy they emit photons with so great a chemical potential that the photon current density emitted toward the Sun has exactly the same magnitude as that obtained from the Sun and no net energy can be transferred to a Carnot engine without the production of entropy.

2.6 Problems

2.1 What is a black body?

2.2 Why does a light-emitting surface of a black body appear equally bright from all viewing angles?

2.3 How many photon states per volume exist in the visible energy range ($\hbar\omega$ between 1.5 and 3 eV)

(a) in vacuum?

(b) in silicon?

(c) At which temperature is the number of photons larger than 10^{-4} of all photon states in this energy range?

2.4 Determine the maximum photon energy $\hbar\omega$ for which the Boltzmann distribution differs from the Bose–Einstein distribution by less than 10^{-2}

(a) for $T = 300$ K

(b) for $T = 5800$ K.

2.5 Calculate the reflectance of a flat Si surface without antireflection coating, using a refractive index of $n_{Si} = 3.5$,

(a) for normal incidence

(b) for incidence at an angle of $\alpha = 50°$ to the surface normal. Assume the light to be unpolarized.

(c) as in case (b), but with light polarized perpendicular to the plane of incidence.

2.6 For normal incidence *AM*1.5 global corresponds to 1000 W m^{-2}.

(a) A solar module is placed on a roof tilted at an angle $\beta = 25°$ (with respect to a horizontal plane) toward the south. At which latitude in the northern hemisphere would the module receive 750 W m^{-2} at noon on a March 21?

(b) Imagine that the module from (a) with an (intensity-independent) efficiency of 15% is installed on a tracking system to gain more power. How many axes are necessary to optimally guide the module on that special day? What is the energy output from the module with and without the tracking system for the March 21 (assume *AM*1.5 global over the entire day)?

2.7 How large is the mean energy of photons from a black body of temperature T absorbed by a semiconductor with an absorptance of $a(\hbar\omega < \varepsilon_G) = 0$ and $a(\hbar\omega \geq \varepsilon_G) = 1$? Assume $\varepsilon_G \gg kT$.

3
Semiconductors

The thermal energy that the absorber in Figure 2.14 supplies to the Carnot engine is contained almost entirely in the random oscillations of the atoms in a solid about their rest positions. The energy of the oscillations is quantized similar to the electromagnetic modes of a cavity. For the electromagnetic field in the cavity, the vibrational quanta are the *photons*, and for the atomic oscillations of a solid, the vibrational quanta are called *phonons*. Phonon energies ε_Γ range from 0 to 0.05 eV. Only in exceptional cases can phonons be directly generated by the absorption of photons. The absorption of photons takes place through the excitation of electrons into states of higher energy. In order that photons of any arbitrary energy can be absorbed, i.e. for a body to be black, a continuous, uninterrupted range of excitation energies must be available to the electrons. For metals, this is in fact the case. Metals would represent the closest approximation to an ideal black body, if they did not reflect most of the incident light. This type of reflection, however, can be eliminated by roughening the surface, which renders a metal completely black. You will experience this yourself when you scrub an aluminum pot with some scouring powder and a cloth. The tiny aluminum particles scrubbed off the pot make the cloth totally black. Owing to the continuous energy range for electrons in a metal, an electron that is excited by absorbing a photon loses this extra energy easily, step by step, in small portions, by generating phonons. Although this may require many steps, the process typically takes place in times of the order of 10^{-12} s. Since in metals, the excitation energy remains for only such a short time in the electron gas, their direct utilization has a poor efficiency, although the emission of the excited electrons out of the metal is used in photomultipliers for the fast detection of single photons. Figure 3.1 is a schematic diagram for the absorption of a photon by an electron in a metal and the subsequent loss of the energy to phonons.

In semiconductors, this mechanism is different. Semiconductors are materials in which the range of excitation energies is interrupted by an energy gap of width ε_G. Figure 3.2 illustrates this schematically. The energy range below the gap, called the *valence band*, is nearly completely occupied with electrons. The energy range above the gap, called the *conduction band* is, however, nearly empty. To excite an electron by the absorption of a photon, the photon must have at least the energy $\hbar\omega = \varepsilon_G$. Photons with smaller energies cannot excite electrons. They are not absorbed. They

Physics of Solar Cells: From Basic Principles to Advanced Concepts. Peter Würfel

ISBN: 978-3-527-40857-3

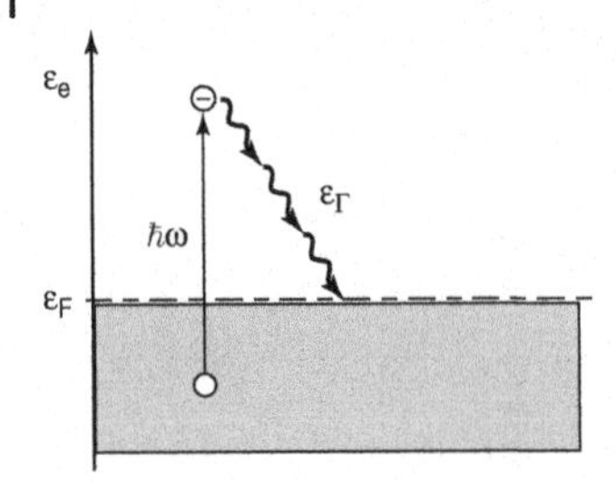

Figure 3.1 Excitation of an electron in the conduction band of a metal by the absorption of a photon with energy $\hbar\omega$ and the subsequent loss of the excitation energy by the generation of single phonons with energy ε_Γ.

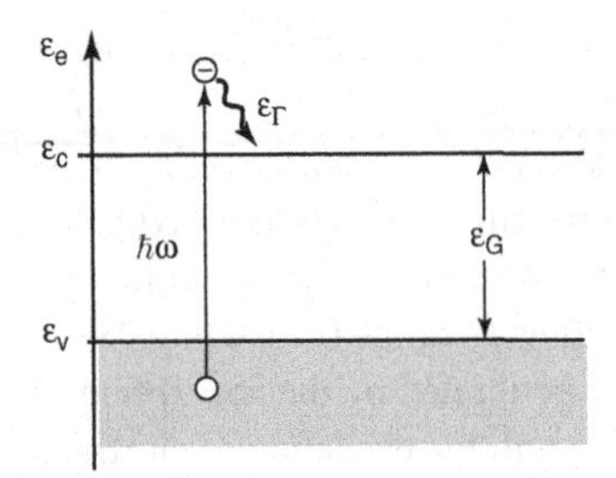

Figure 3.2 Excitation of an electron from the valence band to the conduction band of a semiconductor by the absorption of a photon with energy $\hbar\omega$.

are either transmitted or reflected by the semiconductor. (We ignore the weak absorption by phonons, impurities, and free charge carriers.)

Electrons being excited well into the conduction band in a semiconductor, as shown in Figure 3.2, start losing their energy as quickly as in a metal by stepwise generation of phonons. However, once the electrons have reached the lower edge ε_C of the conduction band, losing energy in small steps by generating phonons is no longer possible, because there are no states for electrons with a little less energy. Returning to a state in the valence band requires the loss of the gap energy ε_G in a single step. Possible processes are the simultaneous generation of a large number of phonons or, alternatively, the emission of a photon. Both processes are, however, much less probable than the stepwise generation of phonons in a continuum of electron states. As a result, the electrons may "live" up to 10^{-3} s in the conduction band. It is this comparatively long time that allows the processes for the conversion of electron energy into electrical energy to take place.

3.1
Electrons in Semiconductors

As we have already seen for photons in a cavity, the density of the electrons dn_e with energy ε_e distributed over the energy interval $d\varepsilon_e$ comprises the density of states $D_e(\varepsilon_e)$ and a distribution function $f_e(\varepsilon_e)$, which defines the occupation of the electron states

$$dn_e(\varepsilon_e) = D_e(\varepsilon_e) f_e(\varepsilon_e)\, d\varepsilon_e \tag{3.1}$$

3.1.1 Distribution Function for Electrons

The distribution of the electrons over the states must satisfy three conditions:

1. According to the Pauli exclusion principle for particles with nonintegral spin, there can never be more than one particle in the same quantum state. This applies for electrons, with a spin $\hbar/2$, in contrast to photons, which have a spin $\hbar$.
2. The occupation of the states depends only on the energy and not, e.g. on the momentum.
3. The occupation of the states must lead to a minimum of the free energy $F = E - TS$.

If the distribution of electrons over the states were to take place in such a way that the energy E is a minimum, the valence band would be completely occupied and the conduction band would be empty for all temperatures T. Under these conditions, the entropy S of the electrons would be zero, since there is only one possibility of producing a fully occupied valence band and an empty conduction band. Promoting a few electrons from the valence band to the conduction band would then lead to not only an increase in the energy, but to a still greater increase in the entropy as well. This is because there are now many possibilities for removing an electron from any of 10^{23} states per cubic centimeter of the valence band and exciting it to any of 10^{23} states per cubic centimeter of the conduction band. The "heat" TS increases and the free energy F is reduced. If there is already a certain number of electrons in the conduction band, the increase in entropy with another electron transition is correspondingly less. The increase in TS exactly compensates for the increase in the energy E when the free energy F reaches its minimum value.

The distribution function that satisfies all of these conditions is the Fermi distribution

$$f_e(\varepsilon_e) = \frac{1}{\exp\left[(\varepsilon_e - \varepsilon_F)/kT\right] + 1} \tag{3.2}$$

This contains the Fermi energy ε_F as a characteristic energy. Figure 3.3 shows the Fermi distribution function. For states with $\varepsilon_e \ll \varepsilon_F$, $f_e(\varepsilon_e) \approx 1$, so that the states are completely occupied. Conversely, for $f_e(\varepsilon_e \gg \varepsilon_F) \approx 0$, states with $\varepsilon_e \gg \varepsilon_F$ are not occupied. Half of the states with $\varepsilon_e = \varepsilon_F$ are occupied.

3.1.2 Density of States $D_e(\varepsilon_e)$ for Electrons

In isolated atoms, the electrons have discrete energy values, separated by large gaps on the energy scale. Decreasing the interatomic distance down to only a

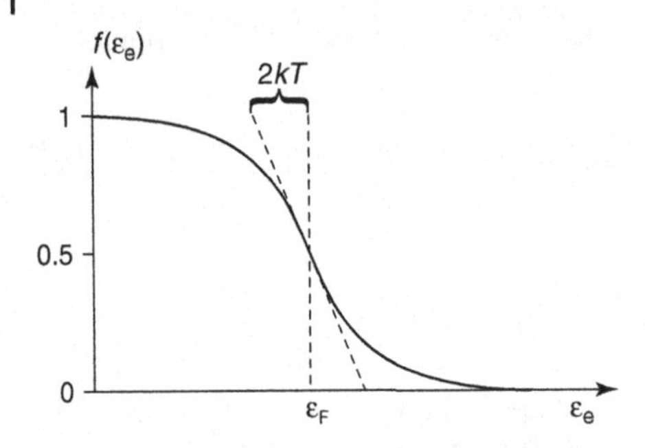

Figure 3.3 The Fermi distribution function $f(\varepsilon_e)$ defines the probability that an electron occupies a state with energy ε_e.

few Å in a solid introduces interaction among the atoms and causes the previously identical energy values to split into as many different values as the solid has atoms. This means that the originally discrete energy values now become energy ranges in which the energy values lie so closely together that they appear to form a continuum. These regions of permitted electron energies are known as *bands*. The stronger the interaction of the electrons with neighboring atoms, the wider the bands become. The interaction is of course strongest for the outer shell valence electrons. Since these electrons have greater energies, the width of the bands increases with increasing energy, while the gaps between the bands decrease correspondingly until they disappear entirely at a certain energy. At higher energies the bands then overlap. Figure 3.4 illustrates this behavior for sodium atoms.

Sodium has the chemical valence one, and correspondingly only one of the two 3s electron states per atom is occupied. The band originating from the 3s states in solid sodium is therefore only half filled with electrons. In the continuous energy range of the 3s band, a great many unoccupied states are thus available to these "valence" electrons. This allows them to be given the additional energy, going along with the necessary additional velocity required for a charge current to flow.

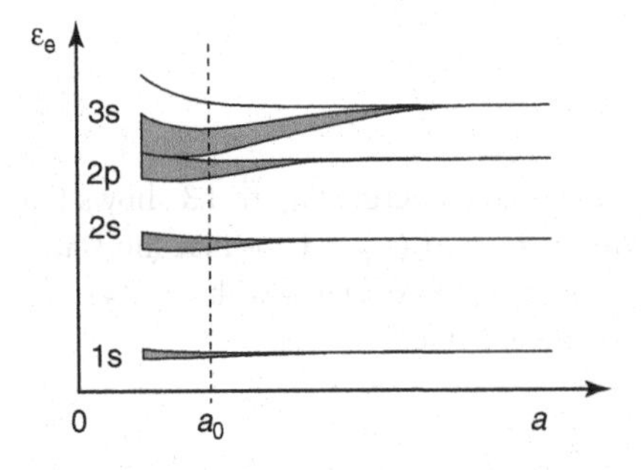

Figure 3.4 Energy of electron states in sodium as a function of the interatomic distance a between the sodium atoms. a_0 is the interatomic distance in solid sodium. States in the shaded energy ranges are occupied by electrons.

A partly occupied band is the prerequisite for metallic conductivity. A partly occupied band is therefore called a *conduction band*. Conversely, pure semiconductors and insulators are nonconductors, because (at least at $T = 0$ K) the uppermost occupied energy band is completely occupied. This is called the *valence band*. The next higher band, separated by the energy gap ε_G, is the conduction band, which (at least at $T = 0$ K) is unoccupied.

To determine the density of states, we proceed exactly as for photons. First, making use of the uncertainty principle, we determine the volume per state and the density of states in momentum space. Then, because the distribution function depends only on the energy, we have to convert the density in momentum space to the density of states per energy interval using the relationship between energy and momentum for electrons.

As for photons, two states differ in location and momentum by at least as much as the uncertainty principle defines, resulting in a phase space volume per state of

$$(\Delta x)^3 (\Delta p)^3 = h^3$$

If we treat the electrons as not localized, their uncertainty in position is $(\Delta x)^3 = V$, where V is the volume of the entire crystal. In momentum space, a state then has the volume

$$(\Delta p)^3 = \frac{h^3}{V}$$

All states with momentum $|p'| \leq |p|$ fill a spherical volume of $(4/3)\pi|p|^3$ in momentum space. The number N of states is found by dividing this volume by the volume of a state in momentum space:

$$N(|p|) = \frac{4\pi|p|^3 V}{3h^3} \tag{3.3}$$

As for photons, these momentum states apply for two electrons with opposite spins. The number of states available for electrons with momentum $|p'| \leq |p|$ is therefore

$$N_e(|p|) = \frac{8\pi|p|^3 V}{3h^3} \tag{3.4}$$

To determine the density of states as a function of the electron energy, we need the relationship between momentum and energy, which for electrons in a crystal might be quite different from that for a free electron. Here we recall that the probability for the occupation of states depends only on their energy. In order that no current flows in the equilibrium state, an equal number of states must be occupied in any arbitrary direction at positive and negative momenta. For this reason, the states must have the same density per energy at positive and negative momenta. Since the states are equidistant in momentum space, we conclude that the energy must be an even function of the momentum. To a first approximation, then

$$\varepsilon_e = \varepsilon_C + \alpha p^2 + \cdots$$

By analogy with free electrons, we set $\alpha = 1/(2m_e^*)$ and call m_e^* the effective mass of the electrons. The kinetic energy of the electrons in the vicinity of ε_C is then given by

$$\varepsilon_{e,\text{kin}} = \varepsilon_e - \varepsilon_C = \frac{p^2}{2m_e^*} \tag{3.5}$$

If the energy depends on the direction of the momentum, as for noncubic crystals, the effective mass m_e^* is a tensor. The energies for oppositely directed momentum values are, however, always identical, so that $\varepsilon_{e,\text{kin}}(p) = \varepsilon_{e,\text{kin}}(-p)$.

Setting p from Equation 3.5 into Equation 3.4, assuming that the effective mass does not depend on the energy, gives the number of states between ε_e and ε_C

$$N_e(\varepsilon_e) = \frac{8\pi V(2m_e^*)^{3/2}}{3h^3}(\varepsilon_e - \varepsilon_C)^{3/2} \tag{3.6}$$

We obtain the density of states D_e in the conduction band as the number of electron states per volume and per energy interval at the energy ε_e by differentiating Equation 3.6.

$$D_e(\varepsilon_e) = \frac{1}{V}\frac{dN_e}{d\varepsilon_e} = 4\pi\left(\frac{2m_e^*}{h^2}\right)^{3/2}(\varepsilon_e - \varepsilon_C)^{1/2} \tag{3.7}$$

Figure 3.5 shows the density of states of the conduction and valence bands for germanium. For the following discussion it is only important that the densities of states at the upper and lower boundaries of a band show square-root behavior as a function of the energy ε_e in agreement with Equation 3.7, provided we choose m_e^* properly. For the upper boundary of a band, at ε_V for the valence band, for example, we must be prepared to choose $m_e^* < 0$. Treating electrons as quasi-free particles although they are in the field of the atoms ($\approx 10^{10}$ Vm^{-1}) will lead to more peculiar properties. For electrons in the conduction band close to ε_C, we actually expect the relationship between energy and momentum shown in Figure 3.6,

$$\varepsilon_e - \varepsilon_C = \frac{p_e^2}{2m_e^*} \tag{3.8}$$

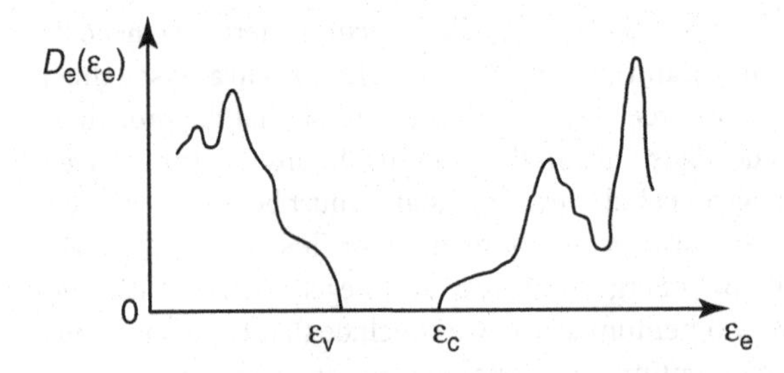

Figure 3.5 Density of states for electrons in the conduction and valence bands of the semiconductor germanium.

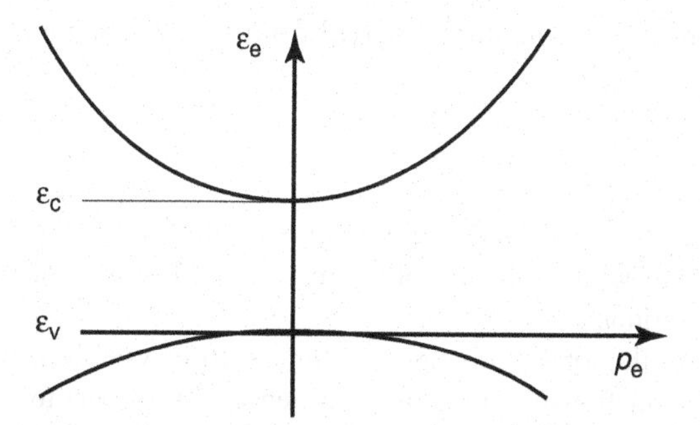

Figure 3.6 Energy ε_e of the electrons for a direct semiconductor as a function of their momentum p_e. The minimum of the conduction band and the maximum of the valence band occur at the same momentum $p_e = 0$.

and obtain the effective mass of the electrons by matching

$$\frac{1}{m_e^*} = \frac{d^2\varepsilon_e}{dp_e^2}$$

to the actual energy–momentum relationship. At the upper boundary ε_V of the valence band $\varepsilon_e - \varepsilon_V < 0$ and according to Equation 3.8 and to $d^2\varepsilon_e/dp_e^2 < 0$, a negative effective mass must be assigned to the electrons.

For the $\varepsilon_e(p_e)$ relationship in Figure 3.6, the excitation of an electron from the valence band to the conduction band with the smallest possible energy, $\varepsilon_C - \varepsilon_V = \varepsilon_G$, occurs without a change in momentum. This excitation is known as a *direct transition,* and semiconductors with this band structure are called *direct semiconductors.* An example of a direct semiconductor is GaAs (gallium arsenide).

Figure 3.7 shows the band structure of an *indirect semiconductor.* This behavior is unexpected from the idea of free particles. Here, the electrons have their lowest

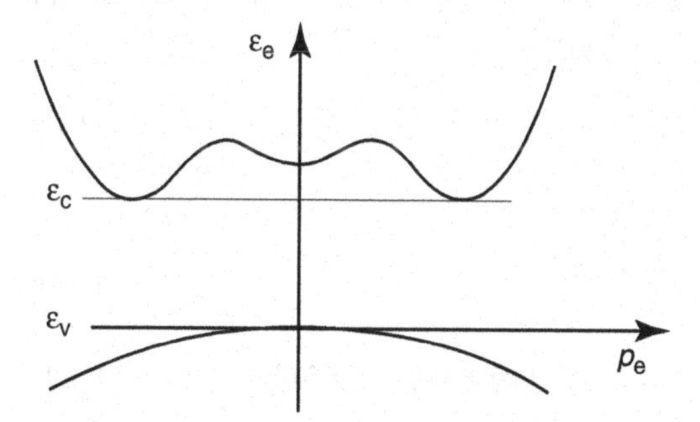

Figure 3.7 Energy of the electrons for an indirect semiconductor as a function of their momentum. The minimum of the conduction band and the maximum of the valence band occur at different values of the momentum p_e of the electrons.

energy ε_C in the conduction band at a momentum different from zero, leading to the relation

$$\varepsilon_e - \varepsilon_C = \frac{(p_e - p_{e,0})^2}{2m_e^*} \tag{3.9}$$

Excitation from the maximum of the valence band to the minimum of the conduction band is only possible with a change of momentum. This type of transition is called an *indirect transition.*

Since in the absence of current the total momentum of the electrons vanishes, the band structure, i.e. the $\varepsilon_e(p_e)$ relationship, must be symmetric with respect to the energy axis at $p_e = 0$. In particular, there must always be an even number of minima for the conduction band when these lie at $p_e \neq 0$. Examples of important indirect semiconductors are germanium (Ge) and silicon (Si).

3.1.3 Density of Electrons

The density of electrons in the energy interval ε_e, $\varepsilon_e + d\varepsilon_e$ is

$$dn_e(\varepsilon_e) = D_e(\varepsilon_e) f_e(\varepsilon_e)\, d\varepsilon_e \tag{3.10}$$

Integration over all energies in the conduction band gives the density of free electrons in the conduction band, or simply the density of electrons. From this point on, the term electrons will refer only to the electrons in the conduction band:

$$n_e = \int_{\varepsilon_C}^{\infty} D_e(\varepsilon_e) f_e(\varepsilon_e)\, d\varepsilon_e \tag{3.11}$$

For the integration, we can make use of the density of states from Equation 3.7, which is, however, valid only for the lower part of the conduction band, and apply it to the entire band (and even beyond this $\to \infty$) because the exponential decrease of $f_e(\varepsilon_e)$ leads to vanishingly small integrands for large values of ε_e. For $\varepsilon_F < \varepsilon_C - 3kT$, we can ignore the "+1" in the denominator of the expression for the Fermi distribution in Equation 3.2, giving

$$n_e = N_C \exp\left(-\frac{\varepsilon_C - \varepsilon_F}{kT}\right) \tag{3.12}$$

with

$$N_C = 2\left(\frac{2\pi m_e^* kT}{h^2}\right)^{3/2} \tag{3.13}$$

N_C is called the *effective density* of states of the conduction band. For $m_e^* = m_e$ and $T = 300$ K, it has the value

$$N_C = 2.51 \times 10^{19}\ \text{cm}^{-3} \tag{3.14}$$

The simplification to the integration obtained by neglecting the "+1" in the denominator of the Fermi function is permissible for $n_e \ll N_C$.

Since a completely occupied valence band does not permit charge transport, the few unoccupied states, known as *holes*, play an important role. Using the same approximation as for electrons, the density of the holes is

$$n_h = \int_{-\infty}^{\varepsilon_V} D_e(\varepsilon_e)\left[1 - f_e(\varepsilon_e)\right] \mathrm{d}\varepsilon_e = N_V \exp\left(-\frac{\varepsilon_F - \varepsilon_V}{kT}\right) \tag{3.15}$$

where $N_V = 2\left(2\pi m_h^* kT/h^2\right)^{3/2}$ is the effective density of states of the valence band containing the effective mass of the holes m_h^* at $\varepsilon_e = \varepsilon_V$. It will be shown in the next section that at $\varepsilon_e = \varepsilon_V$ it is $m_h^* = -m_e^* > 0$ and $\varepsilon_e = -\varepsilon_h$.

Before examining the properties of holes more closely, we will state the important relationship

$$\begin{aligned} n_e n_h &= N_C \exp\left(-\frac{\varepsilon_C - \varepsilon_F}{kT}\right) N_V \exp\left(-\frac{\varepsilon_F - \varepsilon_V}{kT}\right) \\ &= N_C N_V \exp\left(-\frac{\varepsilon_C - \varepsilon_V}{kT}\right) = N_C N_V \exp\left(-\frac{\varepsilon_G}{kT}\right) \end{aligned} \tag{3.16}$$

The product of the electron density and the hole density does not depend on the position of the Fermi energy and therefore it depends neither on the individual densities of the electrons nor on the holes. Consequently, it cannot be influenced by doping. In a pure, so-called intrinsic semiconductor, the electrons in the conduction band originate from the valence band. The density of the electrons n_e is then equal to the density of holes n_h and both are known as *the intrinsic density* n_i.

$$n_e n_h = n_i^2 = N_C N_V \exp\left(-\frac{\varepsilon_G}{kT}\right) \tag{3.17}$$

For an intrinsic semiconductor, the position of the Fermi energy is not exactly in the middle of the energy gap because of different effective densities of states in the conduction and the valence bands. It rather follows from the condition $n_e = n_h$ in Equations 3.12 and 3.15

$$\varepsilon_F = \frac{1}{2}(\varepsilon_V + \varepsilon_C) + \frac{1}{2} kT \ln \frac{N_V}{N_C} \tag{3.18}$$

or expressed in terms of the effective masses contained in N_C and N_V

$$\varepsilon_F = \frac{1}{2}(\varepsilon_V + \varepsilon_C) + \frac{3}{4} kT \ln \frac{m_h^*}{m_e^*} \tag{3.19}$$

Since we know the distribution of the electrons over the states, we can calculate their mean energy

$$\langle \varepsilon_e \rangle = \frac{1}{n_e} \int_{\varepsilon_C}^{\infty} \varepsilon_e D_e(\varepsilon_e) f_e(\varepsilon_e)\, \mathrm{d}\varepsilon_e = \varepsilon_C + \frac{3}{2} kT \tag{3.20}$$

For the electrons, ε_C is a potential energy. The mean kinetic energy of the electrons (in the conduction band) is $\langle \varepsilon_e - \varepsilon_C \rangle = (3/2)\,kT$, showing that the electrons in a nondegenerate semiconductor ($\varepsilon_F < \varepsilon_C - 3kT$), where $n_e \ll N_C$, form an ideal gas.

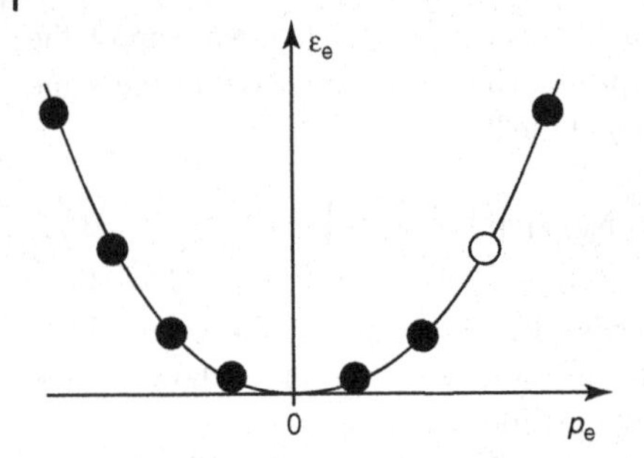

Figure 3.8 A band with a missing electron of $p_e > 0$ has a total momentum $\Sigma p_e < 0$ resulting in an electrical current $j_Q > 0$.

3.2 Holes

For each electron state with momentum p_e, there is also a state with the oppositely directed momentum. For a completely filled band, the total momentum is therefore zero, and no current flows.

We will illustrate the properties of holes for the example of a band with a single unoccupied state. For this purpose, we will first remove an electron with charge $-e$ and velocity $\tilde{v}_e$ from the valence band and secondly, in an alternative description, create the missing electron by adding a hole to the completely filled valence band. The, as yet unknown, properties of the hole will then follow from a comparison of the two descriptions.

The current in a band results from summing over the velocities v_e of all occupied states. The removal of an electron with the velocity $\tilde{v}_e$ from the completely filled band, results in

$$j_Q = -\frac{e}{\mathrm{Vol}} \sum_{\substack{\text{occupied}\\ \text{states}}} v_{e,i} = -\frac{e}{\mathrm{Vol}} \underbrace{\sum_{\text{all states}} v_{e,i}}_{=0\ \text{(filled band)}} - \left(-\frac{e}{\mathrm{Vol}}\tilde{v}_e\right) = \frac{e}{\mathrm{Vol}}\tilde{v}_e \tag{3.21}$$

The same current is obtained by introducing a hole with unknown charge q_h and unknown velocity $\tilde{v}_h$ to a completely filled band:

$$j_Q = -\frac{e}{\mathrm{Vol}} \underbrace{\sum_{\text{all states}} v_{e,i}}_{=0\ \text{(filled band)}} + \underbrace{\left(\frac{q_h}{\mathrm{Vol}}\tilde{v}_h\right)}_{\text{hole}} \tag{3.22}$$

so that

$$j_Q = \frac{e}{\mathrm{Vol}}\tilde{v}_e = \frac{q_h}{\mathrm{Vol}}\tilde{v}_h \tag{3.23}$$

The charge of a band in which an electron is missing is the same as the charge resulting from adding a positive elementary charge to a completely filled band, that is, the charge of the hole is

$$q_h = +e$$

From Equation 3.23 it then follows that $v_h = v_e$ for a hole in a state in which the missing electron has a velocity v_e. This implies that the accelerations of the missing electron (in fact, the sum over all electrons in the band except the missing electron) and of the hole, as a reaction to an electric field E must be identical, so that

$$a_e = -\frac{eE}{m_e^*} = \frac{eE}{m_h^*} = a_h$$

The effective mass of the hole is therefore

$$m_h^* = -m_e^*$$

For the momentum of all electrons of a band in which an electron with momentum $\tilde{p}_e$ is missing, we obtain

$$\bar{p} = \sum_{\text{filled band}} \bar{p}_i - \tilde{p}_e = \sum_{\text{filled band}} \bar{p}_i + \tilde{p}_h$$

$$\tilde{p}_h = -\tilde{p}_e$$

For the energy ε_h of the hole representing the missing electron of energy ε_e, we obtain in the same way

$$\varepsilon_h = -\varepsilon_e$$

The properties of a band result from summing either over all occupied states or over all unoccupied states. For a nearly empty band, such as the conduction band, a description in terms of the occupied states is simple because the few electrons present form an ideal gas. For a nearly filled band, such as the valence band, a description in terms of the occupied states is complicated, because the electrons at the upper edge of the band have a negative effective mass and their velocities and momenta are oppositely directed. A nearly filled band can thus be described more easily in terms of the few unoccupied states, i.e. the holes. These then carry a positive charge, have a positive effective mass, and, as with the electrons in the conduction band, form an ideal gas. Their mean energy is

$$\langle \varepsilon_h \rangle = -\varepsilon_V + \frac{3}{2}kT \tag{3.24}$$

The holes are not merely an imaginary concept, but are just as real as the electrons since the properties of the band can be equally well described in terms of the occupied or the unoccupied states. It is therefore entirely unnecessary first to explain phenomena in which the states of the valence band participate, in terms of electrons, before making the transition to the hole description. The description of holes in the valence band as positively charged particles with a positive effective mass is, in fact, so real that with sufficient kinetic energy they can knock electrons out of their chemical bond, that is, excite them by impact ionization from the valence band to the conduction band.

Figure 3.9 illustrates the transition of an electron from the valence band to the conduction band following the absorption of a photon γ with energy $\hbar\omega$. In this process, the semiconductor absorbs the energy and the momentum of the photon.

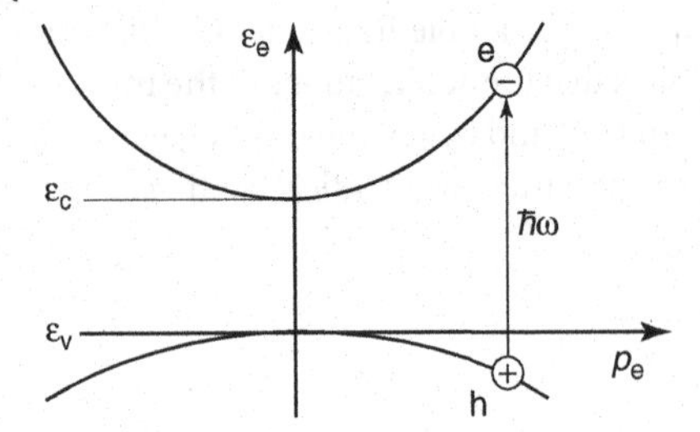

Figure 3.9 Creation of an electron–hole pair by the absorption of a photon.

As a result of the excitation, the conduction band now has an extra electron with momentum $\bar{p}_e$ and energy ε_e, whereas the valence band has an additional hole with momentum $\bar{p}_h$ and energy ε_h. We therefore see the excitation as the creation of an electron (in the conduction band) and a hole (in the valence band) and describe this process as the chemical reaction

$$\gamma \rightarrow e + h$$

This process must of course satisfy the conservation of momentum $\bar{p}_\gamma = \bar{p}_e + \bar{p}_h$ and the conservation of energy $\varepsilon_\gamma = \hbar\omega = \varepsilon_e + \varepsilon_h$.

Figure 3.10 shows an energy scale for electrons. Its zero point is defined by the energy of a free electron in vacuum in an electrical potential $\varphi = 0$ and having a kinetic energy $\varepsilon_{e,\text{kin}} = 0$. In relation to this zero point, the electrons bound in the semiconductor have a negative energy ε_e. The energy ε_h of the holes is then positive. For them the boundaries of the bands ε_C, ε_V should really be drawn folded upward about the zero line. This representation is, however, complicated and is not commonly used. Instead, we enter the hole energies in the same way as the electron energies and take account of their magnitude and sign by the length and direction of arrows. Arrows pointing upward indicate positive energies. For electron energies the arrows point downward, since they are negative with respect to the zero line, and for hole energies the arrows point upward. With this representation,

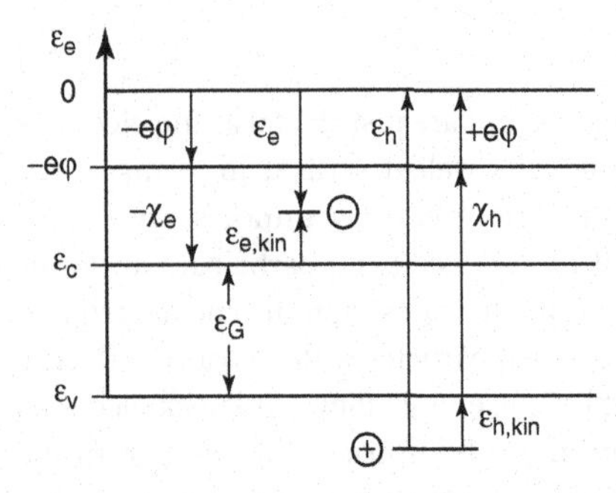

Figure 3.10 Energy scale for electrons and holes in a semiconductor. The binding energy of an electron in a state at the lower boundary of the conduction band is referred to as the *electron affinity* χ_e.

the sum $\varepsilon_e + \varepsilon_h$ is then equal to the difference of the distances from the zero line in Figure 3.10.

3.3 Doping

Doping of semiconductors means the introduction of impurity atoms. In the simplest case, these atoms replace the atoms of the semiconductor at their lattice positions. Figure 3.11 depicts this schematically for a lattice of atoms with a valency of four. Donors (D) are impurity atoms that, as a rule, have more valence electrons than is necessary for chemical bonding with the neighboring atoms. In most practical cases, there is just one electron not required for bonding. It is electrically bound to its atom by Coulomb forces as a negative charge in the field of a positive charge. To a first approximation, we expect this binding energy to be that of the electron in a hydrogen atom with the difference that we have to use the effective mass instead of the real mass of the electron,

$$\varepsilon_H = \frac{m_e^* e^4}{2(4\pi\epsilon_0)^2\hbar^2} = 13.6\ \text{eV}\,\frac{m_e^*}{m_e} \qquad (3.25)$$

Since the donor is, however, in a semiconductor and not in vacuum, the electric field binding the electron to the nucleus is somewhat weakened as a result of polarization of the neighboring atoms. We must therefore replace the dielectric permittivity of free space ϵ_0 in the binding energy of the electron for the H atom by the permittivity of the semiconductor $\epsilon\epsilon_0$. For typical semiconductors such as Ge, Si, and GaAs, the dielectric function ϵ has values >10 for frequencies up to 10^{15} Hz, leading to a drastic reduction in the binding energy down to <0.1 eV. For the same reason, in fact, (the chemical bond of) NaCl dissolves in H_2O.

The electron bound to the donor has the energy ε_D which, because of the weak Coulomb binding, is only slightly smaller than the lower edge ε_C of the conduction band, the lowest energy of free electrons within the semiconductor. Donors therefore donate their electrons easily to the conduction band, a property that gives them their name.

Acceptors (A) are impurity atoms that, as a rule, have one valence electron less than is necessary for chemical bonding with the neighboring atoms. An electron

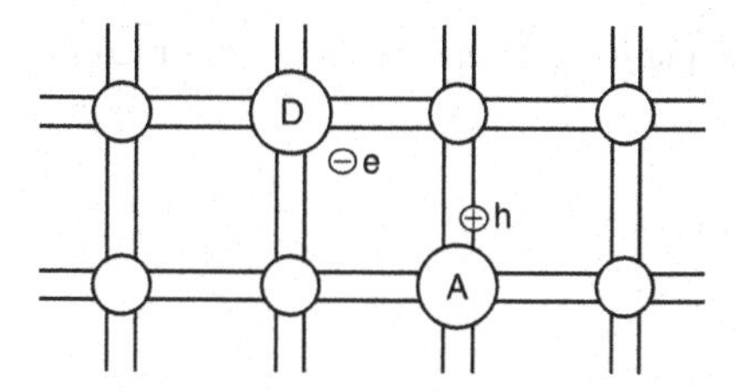

Figure 3.11 By doping lattice atoms are replaced by impurity atoms having a higher (D) or lower (A) valency.

that fills this hole in the bond has no Coulomb attraction to the impurity atom. Consequently, it is not bound as strongly as an electron in the valence band. Since the missing Coulomb bond is, however, only a weak bond, as we have already seen for donors, the energy of an electron at the acceptor is only slightly greater than the upper edge ε_V of the valence band. Acceptors therefore accept an electron easily (i.e. with little excess energy) from the valence band, a property from which their name is derived.

In the hole picture, which, as we have seen, is simpler for the valence band, this means that holes are only weakly bound to acceptors and are easily (i.e. with little expenditure of energy) donated to the valence band.

For lattices of Ge and Si, with a valency of four, P (phosphorus) atoms or As (arsenic) atoms (both with a valency of five) are the most common donors, whereas B (boron) or In (indium) (both with a valency of three) are the usual acceptors.

For the GaAs lattice, consisting of Ga (with a valency of three) and As (with a valency of five), silicon (with a valency of four) is a donor when it is incorporated into a Ga position and an acceptor when it is incorporated into an As position. The type of incorporation in the lattice depends on the temperature. A more important acceptor is Zn (zinc), with a valency of two, at a Ga position, and a more important donor is Cl (chlorine), with a valency of seven, at an As position.

Homogeneous doping is achieved technically by adding the dopant to a semiconductor melt. Inhomogeneous doping, important for semiconductor devices, as we will see, is achieved at high temperatures by the diffusion of dopant atoms from the vapor phase or from the liquid or solid phase in contact with the semiconductor. Another possibility is the implantation of the doping material in ionic form into the semiconductor. With this ion implantation, the spatial distribution of dopant atoms can be more sharply defined. On the other hand, greater damage occurs to the lattice, which may be annealed at high temperatures.

Whether a doping material actually changes the density of electrons or holes depends on the energy of the electrons in the dopant atom, and also on the temperature. For the occupation of all states, the Fermi distribution applies for the bands and the dopant atoms. This defines the density of the positively charged holes and the ionized donors as well as that of the negatively charged electrons and the ionized acceptors. By doping, a semiconductor remains electrically neutral, so that the charge density is

$$\rho = e(n_h + n_D^+ - n_e - n_A^-) = 0 \tag{3.26}$$

Here n_e from Equation 3.12 and n_h from Equation 3.15 are just as much functions of the Fermi energy ε_F as

$$n_D^+ = n_D \left\{ 1 - \left[\exp\left(\frac{\varepsilon_D - \varepsilon_F}{kT} \right) + 1 \right]^{-1} \right\}$$

and

$$n_A^- = n_A \left[\exp\left(\frac{\varepsilon_A - \varepsilon_F}{kT} \right) + 1 \right]^{-1}$$

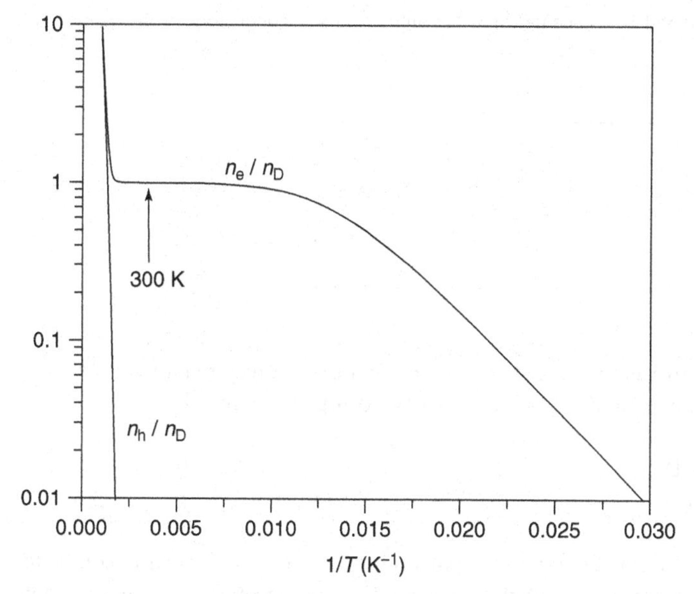

Figure 3.12 Temperature dependence of electron and hole concentrations in an n-type semiconductor with a band gap of 1 eV, containing 10^{16} donors per cm^3 with an energy level $\varepsilon_D = 0.05$ eV below the conduction band edge ε_C.

Equation 3.26 therefore defines the position of the Fermi energy ε_F and with it the values of all the concentrations.

Figure 3.12 shows the temperature dependence of the electron and hole concentrations in an n-type semiconductor with a band gap of 1 eV and effective densities of states of $N_C = N_V = 2 \times 10^{19}$ cm^{-3}, doped with 10^{16} donors per cubic centimeter with an energy level $\varepsilon_D = \varepsilon_C - 0.05$ eV. In a wide temperature range around room temperature, the electron concentration is independent of the temperature and is equal to the donor concentration. According to Equation 3.17, the hole concentration falls off steeply with decreasing temperature. At lower temperatures not all of the donors are ionized. From the decrease of the electron concentration with decreasing temperature, the activation energy for exciting electrons from the donor level to the conduction band (0.05 eV in Figure 3.12) can be obtained. This works, however, only if the donor concentration is much larger than the acceptor concentration. In a more compensated but still n-type semiconductor, the temperature dependence of the electron concentration does not reveal the activation energy of the donors. At high temperatures, electron and hole concentrations rise steeply with increasing temperature. They become equal and much larger than the doping concentration and the semiconductor becomes intrinsic.

As a result of doping with shallow donors to which electrons are only weakly bound, the Fermi energy ε_F at room temperature is less than ε_D, since the donors are unoccupied. For doping with shallow acceptors to which holes are only weakly

Table 3.1 Electron and hole densities in n-type and p-type semiconductors.

	n_e	n_h	ε_F
n-Type	$n_e \approx n_D$	$n_h = \frac{n_i^2}{n_e} = \frac{n_i^2}{n_D}$	$\varepsilon_F = \varepsilon_C - kT \ln \frac{N_C}{n_D}$
p-Type	$n_e = \frac{n_i^2}{n_h} = \frac{n_i^2}{n_A}$	$n_h \approx n_A$	$\varepsilon_F = \varepsilon_V + kT \ln \frac{N_V}{n_A}$

bound, the Fermi energy ε_F at room temperature is greater than ε_A. At room temperature, donors and acceptors are almost completely ionized.

$$\mathrm{D} \longrightarrow \mathrm{D}^+ + \mathrm{e} \qquad\qquad \mathrm{A} \longrightarrow \mathrm{A}^- + \mathrm{h}$$

$$n_e \approx n_D \qquad\qquad n_h \approx n_A$$

With the incorporation of donors, a semiconductor becomes an electron conductor or n-type conductor, and with the incorporation of acceptors it becomes a hole conductor or p-type conductor. Table 3.1 gives the electron and hole densities and the position of the Fermi energy in an n-type and p-type semiconductor doped with either shallow donors or shallow acceptors at about room temperature.

The Fermi energy is obtained from Equation 3.12 and is

$$\varepsilon_F = \varepsilon_C - kT \ln \frac{N_C}{n_e}$$

or from Equation 3.15

$$\varepsilon_F = \varepsilon_V + kT \ln \frac{N_V}{n_h}$$

Typical doping concentrations are in the range of 10^{15} cm^{-3} to 10^{19} cm^{-3}. This is relatively small compared with the lattice atom density, which is about 10^{23} cm^{-3}. This slight admixture does not significantly change the chemical nature of the semiconductor. The energies of the electron and hole states in the bands thus remain largely the same. In order for a slight doping to be effective, the semiconductor must be purified down to concentrations that are small relative to the doping concentrations.

Incorporation of both donors and acceptors in a semiconductor does not lead to increased concentrations of both electrons and holes. Equation 3.17, $n_e n_h = n_i^2$, applies to doped semiconductors as well. The electrons donated from the donors are instead largely taken up by the acceptors. (Alternatively, one could say that the holes that the acceptors leave behind are found in the donor atoms.) Donors and acceptors are both ionized without creating free electrons and free holes. For shallow impurities, only the difference between the donor and acceptor concentrations is effective. Intrinsic concentrations of electrons and holes can therefore be obtained in impure semiconductors as well by compensated doping, making the concentrations of donors and acceptors equal.

Donors have the property that they are neutral with respect to the charge at the unperturbed lattice site when they are occupied by an electron and carry a positive elementary charge when they are unoccupied. Acceptors, on the other hand, have a negative charge in the occupied state and are neutral in the unoccupied state. Donorlike impurities exist with energies not only in the vicinity of the conduction band. For deep donors, the hydrogen atom is not a good model. Deep donors are ineffective as donors, and for solar cells are, in fact, even harmful. If their electron energies are in the middle of the band gap, they act as recombination centers, and if they are close to the valence band they act as hole traps and remove holes from the valence band. Similarly, acceptor-like impurities with electron energies in the middle of the band gap also act as recombination centers, and if they are close to the conduction band they act as electron traps. For good solar cells, in particular, impurities with electron energies in the middle of the band gap must be eliminated to avoid excessive recombination, as discussed in Section 3.6.2.

3.4 Quasi-Fermi Distributions

In a solar cell additional electrons and holes are produced by the absorption of photons from solar radiation. Actually, after their generation, the electron and the hole are bound to each other by Coulomb interaction to form an exciton. The exciton binding energy follows from Equation 3.25 after accounting for the polarizability of the semiconductor lattice by replacing ϵ_0 with $\epsilon\epsilon_0$. Polarization of the lattice and the smaller mass of the hole compared to the mass of the nucleus of an atom reduce the binding energy of the exciton to well below kT_0 (at room temperature). In materials with a large dielectric permittivity like in most inorganic semiconductors, excitons are unstable at room temperature and electrons and holes are essentially free. Figure 3.13 illustrates the generation process schematically. Immediately after the generation of additional electrons and holes, their energy distribution reflects the broad energy spectrum of the absorbed photons (indicating the high photon temperature) and is different from that in the dark state. Owing to collisions with the lattice by which phonons are emitted and absorbed, the energy distribution changes very rapidly. After about 100 collisions over a timescale of 10^{-12} s, the dark-state energy distribution is established where electrons and holes are each in thermal and chemical equilibrium with the phonons and have a mean kinetic energy of $\varepsilon_{e,h,\mathrm{kin}} = (3/2)kT_0$.

After this thermalization, the charge carriers exist for as long as their "lifetime" in their bands, which is long compared with their thermalization time, before they disappear by recombination. Since the thermalization time is so short, the steady-state energy distribution in the bands differs very little from a room temperature distribution, even for continuous irradiation. Owing to the frequent collisions with the atoms, the electrons and holes are in a temperature equilibrium with the

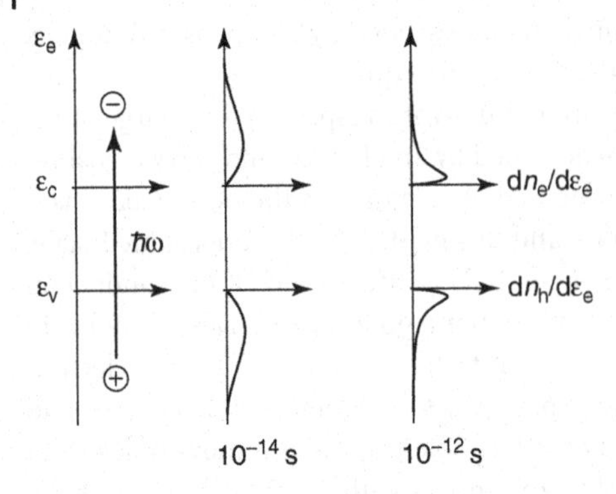

Figure 3.13 The broad energy distribution of electrons and holes, right after their generation by solar radiation, reflects the broad energy spectrum of the photons. After about 10^{-12} s, electrons and holes are thermalized into a narrow room temperature distribution.

lattice, in which Pauli's principle is obeyed and a minimum of the free energy is established, just as in the dark state. The distribution of electrons and holes over the states in their bands must therefore be a Fermi distribution at room temperature.

By irradiation, both the electron and the hole densities are greater than in the dark, that is, $n_e > n_e^0$ and $n_h > n_h^0$, respectively. $n_e n_h$ is then greater than n_i^2, which could not be achieved by doping. The temperature in the Fermi function required for describing the irradiated state must be the lattice temperature, because only this temperature gives a distribution with $\varepsilon_{kin} = (3/2)kT_0$. Owing to the increased electron density, the Fermi energy describing their distribution in the conduction band must be closer to the conduction band than in the dark. Owing to the increased hole density, the Fermi energy describing their distribution in the valence band must be closer to the valence band.

The solution to this dilemma is as follows. There are (always) two Fermi distributions, the distribution f_C with the Fermi energy ε_{FC}, which applies for the occupation of the states in the conduction band and the (shallow) donors with electrons, and another Fermi distribution f_V with the Fermi energy ε_{FV}, which applies for the occupation of the states in the valence band and the (shallow) acceptors with electrons, and therefore also defines the hole density in the valence band.

The density of electrons (in the conduction band) is

$$n_e = N_C \exp\left(-\frac{\varepsilon_C - \varepsilon_{FC}}{kT}\right) \tag{3.27}$$

and that of the holes (in the valence band) is

$$n_h = N_V \exp\left(-\frac{\varepsilon_{FV} - \varepsilon_V}{kT}\right) \tag{3.28}$$

It then follows that

$$n_e n_h = N_C N_V \exp\left(-\frac{\varepsilon_C - \varepsilon_V}{kT}\right) \exp\left(\frac{\varepsilon_{FC} - \varepsilon_{FV}}{kT}\right) \tag{3.29}$$

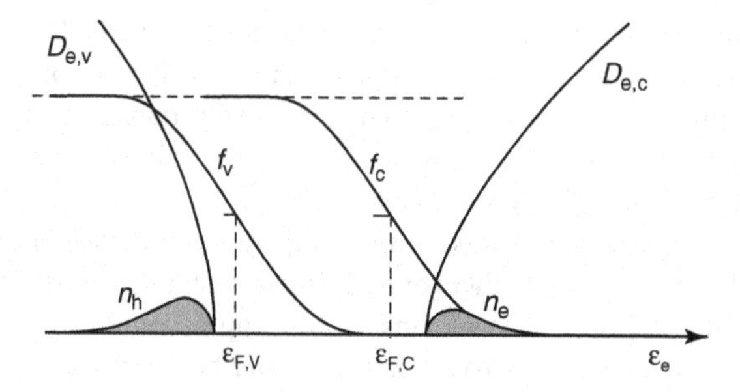

Figure 3.14 In an illuminated semiconductor, the occupations of the conduction band and the valence band are described by different Fermi distributions f_C and f_V.

or

$$n_e n_h = n_i^2 \exp\left(\frac{\varepsilon_{FC} - \varepsilon_{FV}}{kT}\right) \tag{3.30}$$

Figure 3.14 shows these two Fermi distributions. It also shows that especially in the energy range between ε_{FC} and ε_{FV} the distributions contradict each other. If f_C applies, the states in this energy range would have to be occupied, but if f_V applies they would have to be unoccupied. In fact, neither distribution applies in this range of energies. In an illuminated semiconductor the occupation between ε_{FC} and ε_{FV} is governed by kinetics, the preferential capture of holes or electrons. We will return to this point when recombination via impurities is discussed.

3.4.1
Fermi Energy and Electrochemical Potential

On the energy scale shown in Figure 3.10, the energy of the electrons ε_e is divided into potential energy ε_C and kinetic energy ε_{kin}, where ε_C comprises the potential chemical energy $-\chi_e$ also known as the *electron affinity* and the electrical potential energy $-e\varphi$. If an electron is removed from the semiconductor, as is the case when a charge current flows through a solar cell, the energy within the semiconductor decreases by the entire energy ε_e of the electron. But is this energy in fact the energy delivered to a load as electrical energy?

To answer this question we must consider that, according to Gibbs, the exchange of an amount of energy $\mathrm{d}E$ is linked to other quantities that are exchanged as well.

$$\mathrm{d}E(S, V, N_i, Q, \ldots) = T\mathrm{d}S - p\mathrm{d}V + \sum_i \mu_i \mathrm{d}N_i + \varphi\mathrm{d}Q + \cdots \tag{3.31}$$

It is usual to refer to the energy exchange in terms of the quantity exchanged along with the energy. If only entropy S is exchanged with the energy, the energy exchanged is called heat. $-p\mathrm{d}V$ is called *compressional energy*, $\mu_i \mathrm{d}N_i$ is called the *chemical energy* of the particle species i, and $\varphi\mathrm{d}Q$ is called *electrical energy*. There are many other forms of energy as well, e.g. magnetic energy, which are, however, of no interest in connection with solar cells. These energy forms are in all cases products of "intensive" variables and "extensive", quantity-like, variables. To distinguish between these, let us imagine two identical systems, which we put together to form a new system that is twice as large. The intensive variables such as temperature T or pressure p then remain unchanged, while the values of the extensive variables – including the energy – become twice as large.

The extensive quantities, the entropy S, the volume V, the amounts N_i of the different particle species i, the charge Q and others of no importance for solar cells, are the energy carriers. For most of these, laws of conservation apply. For entropy, conservation applies only partially: it can be created, but can never be destroyed.

The intensive variables, temperature T, pressure p, the chemical potential μ_i of the particle species i and electrical potential φ define the amount of energy exchanged with the energy carriers. The gradients of the intensive variables drive currents of the respective energy carriers. Since the current of an energy carrier must vanish when its intensive variable is in equilibrium, the intensive variable then must have the same value everywhere.

When a system is in equilibrium with respect to *one* intensive variable, it is not necessarily in equilibrium with respect to other variables. Figure 3.15 explains this clearly. It shows two compartments of a vessel separated by a mobile piston immobilized for the moment. The two compartments contain different gases: in one there is hydrogen and in the other, oxygen, which therefore have different chemical potentials. In addition, the temperatures and pressures shall also be different. As a result of the temperature difference between the compartments, entropy flows through the still-immobilized piston until the temperatures in the two compartments are the same. Temperature equilibrium, often called simply *thermal equilibrium*, then prevails. However, the pressures and chemical potentials of the gases are still different and not in equilibrium. We will now remove the blockade of the piston, after which it will come to rest (possibly after several oscillations and dissipation of energy) at the position where the pressure is the same in both compartments. We now have equilibrium of temperature and pressure. Chemical equilibrium, that is, the same chemical potentials for hydrogen and oxygen in the two compartments, only occurs by the exchange of gas particles after opening a hole in the piston until each particle type has the same density or concentration

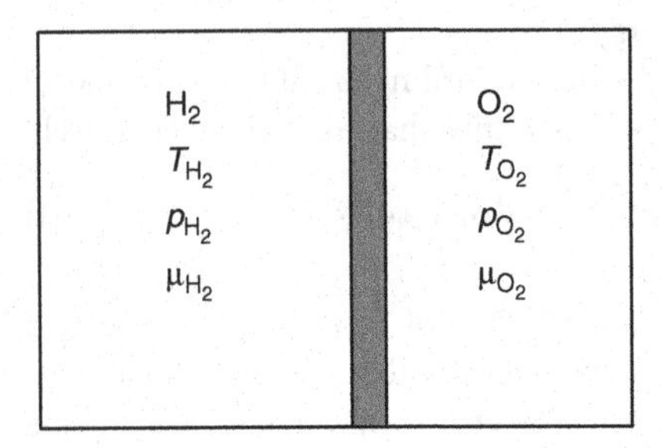

Figure 3.15 Two compartments separated by a piston contain hydrogen H_2 and oxygen O_2 at different temperatures T, pressures p, and chemical potentials μ.

in both compartments. Still another equilibrium, that of the chemical reaction between hydrogen and oxygen to form water, may then be attained by a small spark accompanied by a loud explosion. These equilibria are all independent of each other. That these equilibria never occur all at the same time is of paramount importance for our existence. Consequently, it is entirely meaningful to distinguish between different types of equilibria (and also nonequilibria) and therefore to name these separately as temperature equilibrium, pressure equilibrium, chemical equilibrium, etc. The common designation "thermodynamic equilibrium", while it may sound more impressive, is however meaningless because it says nothing about the type of equilibrium. Furthermore, it also leads us to the false conclusion that "thermodynamic nonequilibrium" prevails and all equilibrium relations are invalid when only one of several intensive quantities is not in equilibrium. Fortunately, a general thermodynamic equilibrium as a state in which all conceivable intensive variables are in equilibrium does not exist in the real world, considering that our very own existence would hardly be possible under such conditions.

In the following, we will make frequent use of the fact that some equilibria exist and allow the use of such equilibrium relations as the Fermi distribution and that there are also nonequilibrium states that are far more difficult to describe.

We will now return to the problem addressed above, namely, the question of how much electrical energy does a semiconductor deliver when an electron–hole pair is removed. Surprisingly, for solar cells just as for batteries, this is not just the energy form $\varphi\,\mathrm{d}Q$, since the charge Q of the solar cell does not change when electrons and holes are removed, again just as for a battery, in which exactly the same number of electrons flows in through one contact as flows out through the other contact. We know, however, that looking for electrical energy in fact means looking for that part of the energy of the electron–hole pairs not accompanied by entropy. This part is the free energy $F(T, V, N_i, Q, \ldots) = E - TS$. With the removal of $\mathrm{d}N_\mathrm{e}$ electrons, the free energy of a body changes by

$$\mathrm{d}F_\mathrm{e}(T, V, N_\mathrm{e}, Q) = \mathrm{d}E_\mathrm{e} - \mathrm{d}(TS_\mathrm{e}) = -S_\mathrm{e}\,\mathrm{d}T - p_\mathrm{e}\,\mathrm{d}V + \mu_\mathrm{e}\,\mathrm{d}N_\mathrm{e} + \varphi\,\mathrm{d}Q$$

With the removal of $\mathrm{d}N_\mathrm{h}$ holes, the free energy changes by

$$\mathrm{d}F_\mathrm{h}(T, V, N_\mathrm{h}, Q) = -S_\mathrm{h}\,\mathrm{d}T - p_\mathrm{h}\,\mathrm{d}V + \mu_\mathrm{h}\mathrm{d}N_\mathrm{h} + \varphi\mathrm{d}Q$$

The total change in the free energy is $\mathrm{d}F = \mathrm{d}F_\mathrm{e} + \mathrm{d}F_\mathrm{h}$

Even if no charge is removed from the solar cell at the end with an electrical current, we must still consider that the electrons and holes are charged particles, so that changes in their numbers are coupled to a change in the total charge.

$$\mathrm{d}Q = z_i\,\mathrm{e}\,\mathrm{d}N_i$$

where $z_i = +1$ for holes and $z_i = -1$ for electrons. It then follows that

$$\mu_i\,\mathrm{d}N_i + \varphi\,\mathrm{d}Q = (\mu_i + z_i e\varphi)\,\mathrm{d}N_i = \eta_i\,\mathrm{d}N_i$$

Owing to this fundamental coupling of charge and particle number, only one coupled electrochemical equilibrium, and no separate chemical and electrical equilibria, exist for the electrons and holes. $\eta_e = \mu_e - e\varphi$ is the electrochemical potential of the electrons. For electrons in electrochemical equilibrium, it has the same value everywhere. $\eta_h = \mu_h + e\varphi$ is the electrochemical potential of the holes and for holes in electrochemical equilibrium, it has the same value everywhere.

For the steady-state operation of solar cells, the temperature T and the volume V occupied by the electrons and holes are constant. Furthermore, with the flow of an electric current, equal numbers of electrons and holes are always removed or added, i.e. $dN_e = dN_h = dN$. The change in the free energy, and thus the energy that a solar cell delivers to a load with dN electrons and holes is then

$$dF = dF_e + dF_h = (\eta_e + \eta_h)\, dN \tag{3.32}$$

The question now remains of how to relate the electrochemical potentials to the already known quantities. In Equation 3.20 we found the mean energy of the electrons to have the value $\langle \varepsilon_e \rangle = \varepsilon_C + (3/2)kT$ and, correspondingly, for the holes, the value $\langle \varepsilon_h \rangle = -\varepsilon_V + (3/2)kT$. We arrive at the same expectation values for the energy per particle of one type by dividing the total energy for particles of one type by the number of such particles. For electrons, the total energy is

$$E_e = TS_e - p_e V_e + \eta_e N_e$$

and the mean energy per electron

$$\frac{E_e}{N_e} = \langle \varepsilon_e \rangle = T\sigma_e - \frac{p_e V_e}{N_e} + \eta_e \stackrel{!}{=} \varepsilon_C + \frac{3}{2}kT \tag{3.33}$$

Similarly, for holes

$$\frac{E_h}{N_h} = \langle \varepsilon_h \rangle = T\sigma_h - \frac{p_h V_h}{N_h} + \eta_h \stackrel{!}{=} -\varepsilon_V + \frac{3}{2}kT \tag{3.34}$$

The quantities σ_e and σ_h are the entropy per electron and per hole, respectively. Since electrons and holes are ideal gases, we can make use of the relationship between σ and the particle density n discovered by Sackur and Tetrode for ideal gases, after adapting it to particles with spin 1/2

$$\sigma = k\left\{\frac{5}{2} + \ln\left[\frac{2\left(\frac{2\pi mkT}{h^2}\right)^{3/2}}{n}\right]\right\} \tag{3.35}$$

Substituting the effective mass m^* of the electrons and holes from Equation 3.13 for m we find

$$\sigma_{e,h} = k\left(\frac{5}{2} + \ln\frac{N_{C,V}}{n_{e,h}}\right) \tag{3.36}$$

For ideal gases with particle number N the equation of state is

$$pV = NkT \tag{3.37}$$

With Equations 3.36 and 3.37, we find from Equation 3.33 for electrons

$$\langle \varepsilon_e \rangle = \varepsilon_C + \frac{3}{2}kT = kT\left[\frac{5}{2} + \ln\left(\frac{N_C}{n_e}\right)\right] - kT + \eta_e$$

$$\varepsilon_C - \eta_e = kT\ln\left(\frac{N_C}{n_e}\right)$$

which leads to

$$n_e = N_C \exp\left[\frac{-(\varepsilon_C - \eta_e)}{kT}\right] \tag{3.38}$$

This is identical with Equation 3.27, and we can now identify the electrochemical potential of the electrons with their Fermi energy

$$\eta_e = \varepsilon_{FC} \tag{3.39}$$

For holes a similar result applies; thus

$$-\varepsilon_V - \eta_h = kT\ln\left(\frac{N_V}{n_h}\right) \tag{3.40}$$

and

$$n_h = N_V \exp\left(\frac{\eta_h + \varepsilon_V}{kT}\right) \tag{3.41}$$

Comparison with Equation 3.28 shows that

$$\eta_h = -\varepsilon_{FV} \tag{3.42}$$

These values are shown in Figure 3.16. The free energy delivered to a load by dN electrons and holes is

$$dF = dF_e + dF_h = (\eta_e + \eta_h)\, dN = (\varepsilon_{FC} - \varepsilon_{FV})\, dN$$

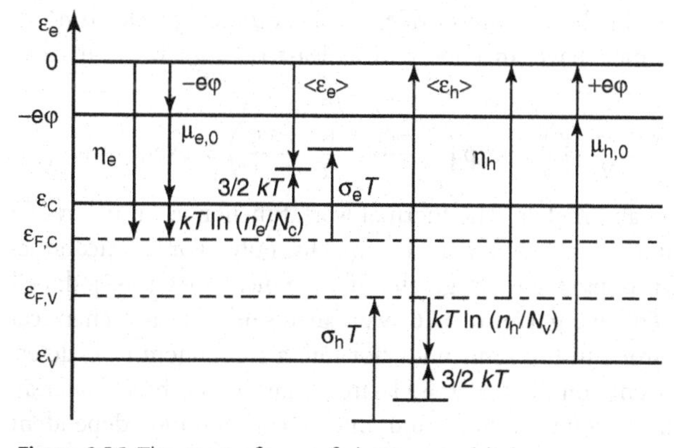

Figure 3.16 The energy forms of electrons and holes.

This result is in good agreement with the expectation that the electron–hole system of an unilluminated semiconductor, that is, one that sees only the 300 K ambient radiation and is in equilibrium with it, cannot deliver electrical energy because for this state $\varepsilon_{FC} - \varepsilon_{FV} = \eta_e + \eta_h = 0$. The electrochemical energy of an electron–hole pair, being the sum of the electrochemical potentials of an electron and a hole at the same location x at which the electron and the hole are in the same electrical potential φ, is equal to the chemical energy of the electron–hole pair because an electron–hole pair is uncharged:

$$\eta_e(x) + \eta_h(x) = \mu_e(x) - e\varphi(x) + \mu_h(x) + e\varphi(x) = \mu_e(x) + \mu_h(x) \qquad (3.43)$$

This is important for the conversion of solar energy into chemical energy as in photosynthesis. This conversion takes place in every semiconductor exposed to light without any additional measures.

3.4.2 Work Function

The absolute value of the chemical potential for electrons μ_e is also known as the *work function*. This is the energy that must be expended in order to excite an electron from a bound state in a semiconductor or metal thermally, i.e. by thermionic emission, to a state of a free electron in vacuum. We assume that while leaving the solid the electron is still in the same electrical potential so that its electrical energy $-e\varphi$ remains unchanged. For metals, in which all states up to the Fermi energy are occupied, it is evident from Figure 3.16 that the work function is equal in magnitude to the chemical potential μ_e. In semiconductors, however, there are no electrons at all with energies equal to the Fermi energy, since the Fermi energy is somewhere within the band gap.

The work function ϕ is determined experimentally from the temperature dependence of the electron current emitted by thermionic emission, which is found to be proportional to $\exp(-\phi/kT)$. This emission current is proportional to the concentration of electrons $n_{e,free}$ not bound to the semiconductor material, but still belonging to the same Fermi distribution, which have an energy of at least $\varepsilon_e = -e\varphi$. As in the derivation of the concentration of electrons in the conduction band, which have an energy of at least $\varepsilon_e = \varepsilon_C$, we find from Equation 3.12

$$n_{e,\,free} \propto \exp\left(-\frac{-e\varphi - \varepsilon_F}{kT}\right) = \exp\left(-\frac{-e\varphi - \mu_e + e\varphi}{kT}\right) = \exp\left(\frac{\mu_e}{kT}\right)$$

for both semiconductors and metals. The thermal work function as an invariable property of a compound is characteristic for metals only. For semiconductors, it depends on the doping and is greater if a semiconductor is p-doped than if it is n-doped. Following Figure 3.16, we can decompose the chemical potential μ_e of the electrons in the semiconductor into a component $\mu_{e,0}$, determined by the chemical environment of the electrons, i.e. by the base material, and independent of their concentration, and into a concentration-dependent component

$$\mu_e = \mu_{e,0} + kT \ln\left(\frac{n_e}{N_C}\right) \tag{3.44}$$

For semiconductors, the property that is characteristic for the base material is the concentration-independent component of the chemical potential of the electrons $\mu_{e,0}$ or its magnitude, the electron affinity $\chi_e = -\mu_{e,0}$, from which the work function is calculated according to Equation 3.44 for a known electron concentration.

Work functions can also be obtained experimentally from photoemission as the smallest photon energy for which the emission of electrons into the vacuum is observed. For metals, there is no significant difference compared with the thermionic work function. For semiconductors, however, photoemission does not measure the work function but the energy of the transition from the upper edge of the valence band into the vacuum. Owing to the much lower density of electrons in the conduction band, transitions from the conduction band into the vacuum are much more difficult to observe.

3.5 Generation of Electrons and Holes

Electrons and holes are produced by processes that can supply at least the minimum generation energy ε_G of an electron–hole pair. This includes impact ionization, in which an electron (or hole) with enough kinetic energy knocks a bound electron out of its bound state (in the valence band) and promotes it to a state in the conduction band, thereby creating an electron and a hole. The same process of exciting an electron from the valence band to the conduction band may take place with a giant lattice vibration supplying the energy or by the absorption of a photon. In the presence of impurities providing states with energies in the energy gap, the excitation can take place in several steps and the generation energy ε_G can be supplied in smaller portions by photons or even phonons. In all cases, in close proximity the electron and hole attract each other by their opposite charge. If their energy is insufficient to overcome the attraction, a bound state of electron and hole results, which is called an *exciton*. In inorganic semiconductors, the exciton binding energy is smaller than kT at room temperature and electrons and holes are essentially free particles. In contrast, organic semiconductors usually have exciton binding energies much larger than kT. The properties of the exciton and the problems that a large exciton binding energy presents for organic solar cells will be discussed in Section 6.9.

3.5.1 Absorption of Photons

For solar cells, the generation of electrons and holes by the absorption of photons is the most important process. The probability for the absorption of a photon of energy $\hbar\omega$ is defined by the absorption coefficient $\alpha(\hbar\omega)$, which is a material property, independent of the geometry of a body. Since absorption requires that an electron–hole pair is generated, $\alpha(\hbar\omega)$ is proportional to the density of occupied states in the valence band in which a hole can be generated, and unoccupied states in the conduction band in which an electron can be generated. The changes in occupation by the absorption of a photon must, of course, conserve momentum and energy.

Direct Transitions

Direct transitions are those in which the momentum of the electron–hole system does not change. The balance of momentum is then consistent with a reaction exclusively with photons, which have a negligible momentum as a result of the very large value of the velocity of light in the definition of their momentum $p_\gamma = \hbar\omega/c$,

$$p_\gamma = p_\mathrm{e} + p_\mathrm{h} \approx 0, \quad \text{and thus} \quad p_\mathrm{e} = -p_\mathrm{h}$$

Energy conservation, $\hbar\omega = \varepsilon_\mathrm{e} + \varepsilon_\mathrm{h}$ together with the energy–momentum relations of direct semiconductors shown in Figure 3.6,

$$\varepsilon_\mathrm{e} = \varepsilon_\mathrm{C} + \frac{p_\mathrm{e}^2}{2m_\mathrm{e}^*} \quad \text{and} \quad \varepsilon_\mathrm{h} = -\varepsilon_\mathrm{V} + \frac{p_\mathrm{h}^2}{2m_\mathrm{h}^*} \tag{3.45}$$

leads to

$$\hbar\omega = \varepsilon_\mathrm{C} - \varepsilon_\mathrm{V} + \frac{p_\mathrm{e}^2}{2m_\mathrm{e}^*} + \frac{p_\mathrm{h}^2}{2m_\mathrm{h}^*}$$

Since from the conservation of momentum $p_\mathrm{e}^2 = p_\mathrm{h}^2 = p^2$, it follows that

$$\hbar\omega = \varepsilon_\mathrm{G} + \frac{p^2}{2}\left(\frac{1}{m_\mathrm{e}^*} + \frac{1}{m_\mathrm{h}^*}\right) = \varepsilon_\mathrm{G} + \frac{p^2}{2m_\mathrm{comb}} \tag{3.46}$$

$m_\mathrm{comb} = (m_\mathrm{e}^* m_\mathrm{h}^*)/(m_\mathrm{e}^* + m_\mathrm{h}^*)$ is the so-called combined mass.

In Equation 3.46 the dependence of the energy $\hbar\omega$ of a direct transition on the momentum is very similar to that of the energy–momentum relation of an electron in the conduction band (Equation 3.45). In the same way as the quantization of momentum together with this relation leads to the density of electron states in the conduction band in Equation 3.7, the energy–momentum relation in Equation 3.46 results in the so-called combined density of states for direct optical transitions,

$$D_\mathrm{comb}(\hbar\omega) = \frac{4\pi}{h^3}(2m_\mathrm{comb})^{3/2}(\hbar\omega - \varepsilon_\mathrm{G})^{1/2} \tag{3.47}$$

The probability of photon absorption, i.e. the change in the photon current density $\mathrm{d}j_\gamma$ over a distance $\mathrm{d}x$ by absorption is proportional to the absorption coefficient and to j_γ. This defines the absorption coefficient $\alpha(\hbar\omega)$, which is proportional to the combined density of states,

$$\frac{\mathrm{d}j_\gamma(\hbar\omega)}{\mathrm{d}x} = -\alpha(\hbar\omega)\, j_\gamma(\hbar\omega) \tag{3.48}$$

$$\alpha(\hbar\omega) \propto D_\mathrm{comb}(\hbar\omega) \propto (\hbar\omega - \varepsilon_\mathrm{G})^{1/2} \tag{3.49}$$

Integration of Equation 3.48 leads to the well-known dependence of the photon current on the distance x from the surface at $x = 0$ in an absorbing body:

$$j_\gamma(x) = j_\gamma(0)\exp(-\alpha x)$$

For semiconductors with the energy–momentum relation shown in Figure 3.6, in which direct transitions with almost no change in the momentum are possible we expect the square-root dependence on energy for the absorption coefficient. Figure 3.17 gives the absorption coefficient α of GaAs as an example.

Photons with $\hbar\omega < \varepsilon_G$ are not absorbed by electron–hole generation and $\alpha(\hbar\omega < \varepsilon_G) = 0$. Photons of this energy are either reflected or transmitted. For $\hbar\omega > \varepsilon_G$, α shows the steep rise in accordance with the theoretical expectations $\propto (\hbar\omega - \varepsilon_G)^{1/2}$ up to values of 10^4 cm^{-1}. For larger values of α the complexity of the real band structure of GaAs causes deviations from the square-root dependence. At the absorption edge at $\hbar\omega = \varepsilon_G = 1.4$ eV, an exponential increase is superimposed over the square-root dependence. This is known as the *Urbach tail*, which results from statistical fluctuations of the band gap caused by lattice vibrations. Since for a distance from the surface $x = 1/\alpha$ the intensity is attenuated by a factor of e, $L_\gamma = 1/\alpha$ is called the *penetration depth of the photons*. Owing to the large absorption coefficient and the small penetration depth of the photons in GaAs and other direct semiconductors, a solar cell made from these materials does not have to be thicker than only a few μm in order to absorb the absorbable part of the solar spectrum with photon energies $\hbar\omega > \varepsilon_G$.

Indirect Transitions

In an indirect semiconductor, a transition between the maximum of the valence band ε_V and the minimum of the conduction band ε_C is not possible with only the absorption of a photon, because the momentum $p_\gamma = \hbar\omega/c$ of the photon is too small. The momentum balance is satisfied through the participation of another "particle", a lattice vibration or a phonon. Due to the large mass of the atoms, phonons with small energies $\hbar\Omega$ have a large momentum p_Γ.

In an indirect transition a phonon Γ can either be absorbed with the absorption of a photon γ,

$$\begin{aligned} \gamma + \Gamma &\rightarrow \mathrm{e} + \mathrm{h} \\ p_\gamma + p_\Gamma &= p_\mathrm{e} + p_\mathrm{h} \\ \hbar\omega + \hbar\Omega &= \varepsilon_\mathrm{e} + \varepsilon_\mathrm{h} \end{aligned} \tag{3.50}$$

or the phonon Γ can be emitted

$$\begin{aligned} \gamma &\rightarrow \mathrm{e} + \mathrm{h} + \Gamma \\ p_\gamma &= p_\mathrm{e} + p_\mathrm{h} + p_\Gamma \end{aligned}$$

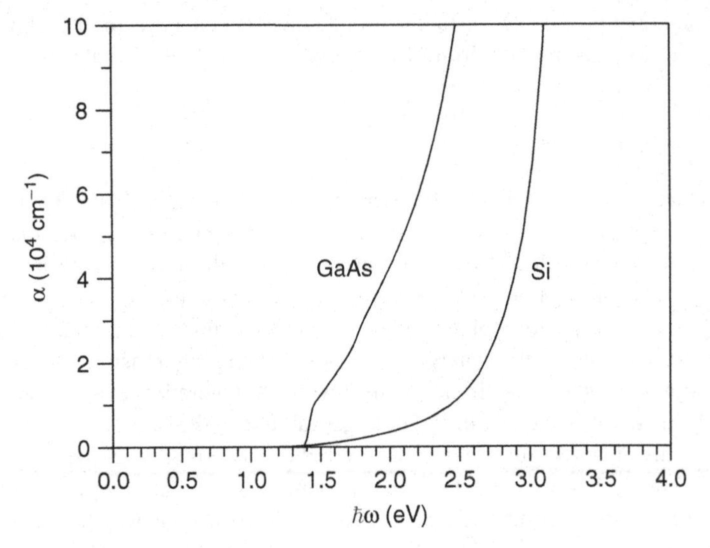

Figure 3.17 Absorption coefficient α of the "direct" semiconductor gallium arsenide and the "indirect" semiconductor silicon.

$$\hbar\omega = \varepsilon_e + \varepsilon_h + \hbar\Omega \tag{3.51}$$

The participation of the phonons allows photon-induced transitions from every state of the valence band to every state of the conduction band for which the energy balance shown in Figure 3.18 is satisfied. The transition probability into *one* state of the conduction band is then proportional to the number of all states in the valence band separated from this conduction band state by the photon energy $\hbar\omega$ plus or minus the phonon energy $\hbar\Omega$. To find the probability for the absorption of a photon of energy $\hbar\omega$ we have to integrate over all states of the conduction band with the energy $\varepsilon_{e,\,kin}$ that are accessible from the valence band, with the simultaneous emission or absorption of a phonon. The smallest value for the kinetic energy of the electrons is $\varepsilon_{e,\,kin} = 0$ for a transition to the minimum of the conduction band and the highest possible value is $\varepsilon_{e,\,kin} = \hbar\omega \pm \hbar\Omega - \varepsilon_G$ for a transition from the maximum of the valence band to the conduction band.

$$\alpha(\hbar\omega) \propto \int_0^{\hbar\omega \pm \hbar\Omega - \varepsilon_G} D_C(\varepsilon_{e,\,kin}) D_V(\hbar\omega \pm \hbar\Omega - \varepsilon_G - \varepsilon_{e,\,kin})\, d\varepsilon_{e,\,kin}$$

With the energy dependence of the densities of states of Equation 3.7 the integration yields

$$\alpha(\hbar\omega) \propto (\hbar\omega - \varepsilon_G \pm \hbar\Omega)^2 \tag{3.52}$$

The plus sign applies for the simultaneous absorption of a photon and a phonon and the minus sign for the emission of a phonon with the absorption of a photon. Because of the requirement of phonon participation in order to satisfy the conservation of momentum, the absorption coefficient of an "indirect" semiconductor is

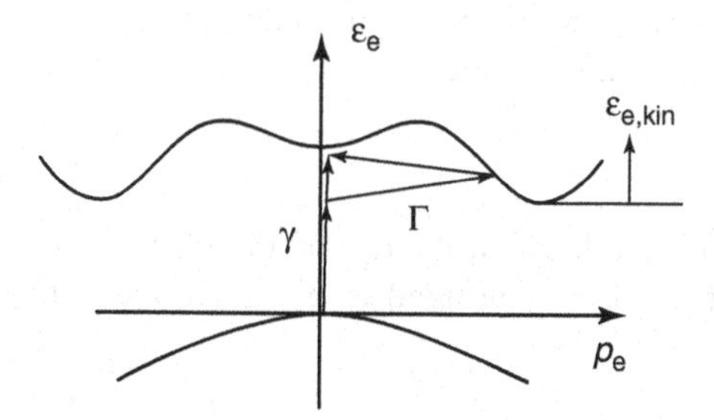

Figure 3.18 Energy ε_e of electron states in the conduction and valence bands between which transitions are possible with the absorption of a photon γ and the simultaneous absorption or emission of a phonon Γ.

small. Figure 3.17 shows the dependence of the absorption coefficient on the photon energy for silicon, a typical indirect semiconductor. The quadratic dependence on the photon energy is well developed. Near the band gap $\hbar\omega = \varepsilon_G = 1.12$ eV the absorption coefficient of silicon is so small that the band gap cannot be seen in Figure 3.17. Also, the absorption and emission of phonons cannot be distinguished.

Because of the small value of α for indirect transitions, the penetration depth for photons in an indirect semiconductor is large. To absorb all the photons with $\hbar\omega > \varepsilon_G$, which can be absorbed from the solar spectrum, an indirect semiconductor in the geometry of a plane-parallel plate must be more than 100 μm thick. Along with silicon, germanium is also an indirect semiconductor.

3.5.2
Generation of Electron–Hole Pairs

An important equation for the following considerations is the continuity equation. In its general form for a quantifiable variable, e.g. for the concentration of a certain type i of particle, it is

$$\frac{\partial n_i(x)}{\partial t} = G_i(x) - R_i(x) - \operatorname{div} j_i(x) \tag{3.53}$$

This relation expresses the simple consideration that the density n_i of particles of type i in a volume element at the location x increases with time when these particles are generated at a rate G_i, and that n_i decreases when particles at the location x are annihilated at the rate R_i or when they escape from the volume element at the location x because the particle current density j_i flowing away toward the right (to larger x) is greater than that flowing in from the left (from smaller x), expressed by $\operatorname{div} j_i > 0$.

In order to gain familiarity with the use of the continuity equation, we will apply this equation to a case for which we already know the solution, namely, to the photon current penetrating the semiconductor and being attenuated as a result of absorption. If an external photon current density j_γ is incident onto the semiconductor and no photons are created within the semiconductor ($G_\gamma = 0$), then in steady state, where no changes occur over time,

$$\frac{\partial n_\gamma}{\partial t} = -R_\gamma - \operatorname{div} j_\gamma = 0$$

With $R_\gamma = \alpha j_\gamma$ the expected proportionality between the rate of annihilation and the photon density (which moves at the velocity of light) is expressed. For light incident in the x direction this yields

$$\operatorname{div} j_\gamma = \frac{\mathrm{d}j_\gamma}{\mathrm{d}x} = -R_\gamma = -\alpha j_\gamma$$

Integrating, we find

$$j_\gamma(x) = j_\gamma(0)\exp(-\alpha x)$$

the absorption law with which we are already familiar. Here, $j_\gamma(0)$ is the photon current density penetrating the surface of the semiconductor. It is less than the incident photon current density $j_{\gamma,\mathrm{in}}$ due to reflection at the surface

$$j_\gamma(0) = (1 - r)\, j_{\gamma,\mathrm{in}}$$

With the transmittance of a body of thickness d

$$t = (1 - r)\exp(-\alpha d)$$

the total absorbed photon current becomes

$$j_{\gamma,\mathrm{abs}} = (1 - r - t)\, j_{\gamma,\mathrm{in}} = (1 - r)\,[1 - \exp(-\alpha d)]\, j_{\gamma,\mathrm{in}} = a\, j_{\gamma,\mathrm{in}}$$

The absorptance $a = (1 - r)[1 - \exp(-\alpha d)]$ applies for a plane-parallel plate of thickness d, if multiple reflection is neglected.

Since both a hole and an electron are produced for each absorbed (or annihilated) photon, they are both generated at the rate

$$G_\mathrm{h} = G_\mathrm{e} = R_\gamma = \alpha\, j_\gamma(x)$$

This assumes that an absorbed photon always generates only one electron and one hole. However, if the photon energy $\hbar\omega$ is at least twice as large as the band gap ε_G, one of the two charge carriers can then have a kinetic energy of $\varepsilon_\mathrm{kin} \geq \varepsilon_\mathrm{G}$. This charge carrier, whether an electron or a hole, is then able to knock an electron out of its chemical bond and thus produce another electron–hole pair by impact ionization. Owing to the symmetry of the band structure of the valence and conduction band, however, the energy of the photon is more or less equally divided among the electron and the hole. In this case multiple electron–hole pair generation by impact ionization would occur only for photon energies $\hbar\omega > 3\varepsilon_\mathrm{G}$. This fundamental possibility requires a band structure for the semiconductor that ensures the conservation of energy and momentum. In real solar cells with band gaps ≥ 1 eV, however, it occurs only with very low probability and requires photon energies hardly found in the solar spectrum.

As an example for the production of one electron–hole pair per absorbed photon, we will calculate the generation rate of electrons and holes resulting from the absorption of the 300 K black-body radiation of the environment. As equilibrium rates they will be labeled with a superscript 0. Photons of different energies contribute according to the value of the absorption coefficient $\alpha(\hbar\omega)$:

$$G_e^0 = G_h^0 = R_\gamma^0 = \int_0^\infty \alpha(\hbar\omega)\,\mathrm{d}j_\gamma(\hbar\omega)$$

$$= \frac{\Omega}{4\pi^3\hbar^3c^2}\int_0^\infty \frac{\alpha(\hbar\omega)(\hbar\omega)^2}{\exp\left(\dfrac{\hbar\omega}{kT_0}\right)-1}\,\mathrm{d}\hbar\omega \qquad (3.54)$$

c is the speed of propagation in the medium and is smaller than in vacuum by the refractive index n of the medium. This means that the photon current density per solid angle, and therefore the absorption rate, is proportional to n^2. Since the photon current density does not change during the transition from the vacuum into the medium except by reflection, the larger photon current density per solid angle in the medium must be restricted to a range of solid angles into which photons flow from outside, which is smaller by a factor of n^2 than the solid angle range Ω_{vacuum} in the vacuum from which they come. This is the reason for Snell's law of refraction, according to which the photons make a smaller angle with respect to the surface normal in a medium than in vacuum. This is illustrated in Figure 3.19.

The generation rate of electron–hole pairs by photons incident from outside follows from Equation 3.54 by setting $\Omega = \pi$ and $c = c_0$, the vacuum velocity of light. These are, however, not all photons absorbed in the medium. Let us assume that the semiconductor is in thermal and chemical equilibrium with its environment and that no current flows. Then at every location within the semiconductor and in

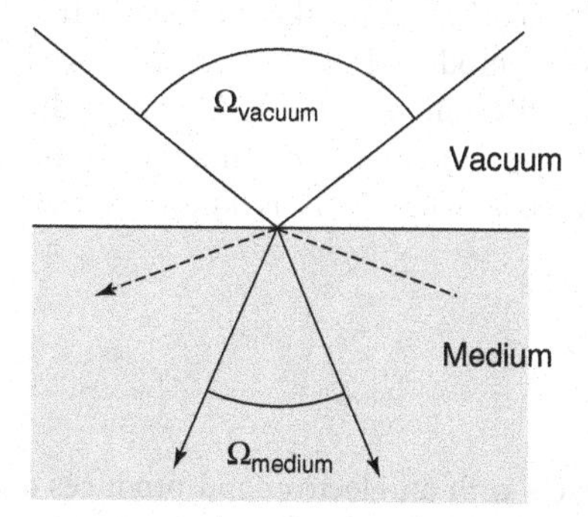

Figure 3.19 In accordance with the law of refraction at the interface between two media, only photons from a smaller solid angle in the medium with the larger refractive index are exchanged with the other medium. Photons for which the momentum lies in the solid angle range inaccessible from outside undergo total internal reflection (dashed light path).

every solid angle element, exactly the same number of photons must be emitted as are absorbed. Since photon emission is generally isotropic, the radiation in the medium is also isotropic with $\Omega_{medium} = 4\pi$. Furthermore, in this state of equilibrium the radiation in the semiconductor is also homogeneous, so that $j_\gamma^0 \neq j_\gamma^0(x)$. As we have just seen, certain solid angle ranges are not accessible to photons from outside. According to the principle of detailed balance the photons emitted inside the medium into the inaccessible solid angle range cannot leave the medium; they undergo total internal reflection. In equilibrium with the 300 K environment, these photons are absorbed and reemitted inside the medium and contribute to the generation and recombination rates, which then follow from Equation 3.54 by setting $\Omega = 4\pi$ and $c = c_0/n$, the speed of light in the medium.

For electrons and holes it is not meaningful to define a density per volume and per solid angle, because after a few collisions with phonons, i.e. in less than 10^{-13} s, an isotropic momentum distribution is established. The momentum of electrons and holes is then uniformly distributed over a solid angle range $\Omega = 4\pi$.

In addition to the generation of electrons and holes by the absorption of photons, they may also be generated by thermal transitions from the valence band to the conduction band in the absence of radiation, involving impurities. In thermal and chemical equilibrium with the environment, the nonradiative generation and recombination rates must also be balanced. We will come back to thermal generation of electron–hole pairs by the absorption of phonons in the next section, where we discuss nonradiative recombination.

3.6
Recombination of Electrons and Holes

The different processes for the production of electrons and holes discussed in Section 3.5 also exist in reverse, that is, processes in which electrons and holes are annihilated in a reaction known as *recombination*. From the energy set free in this reaction, either photons or phonons are produced or both are produced simultaneously. In thermal and chemical equilibrium with the radiation in the environment, in which $n_e^0 n_h^0 = n_i^2$, the rates of production and annihilation are exactly balanced for each of the different mechanisms. This is known as the *principle of detailed balance.*

3.6.1
Radiative Recombination, Emission of Photons

Radiative recombination, in which a hole reacts with an electron and produces a photon, is exactly the reverse of absorption (in the pure electron description it is the spontaneous transition of an electron from the conduction band to an unoccupied state in the valence band).

$$\mathrm{e} + \mathrm{h} \longrightarrow \gamma$$

Since a free electron and a free hole must find each other, the rate of radiative recombination at which electrons and holes are annihilated and photons are generated increases with the concentration of electrons and the concentration of holes

$$G_\gamma = R_e = R_h = Bn_e n_h \tag{3.55}$$

Here B is the coefficient for radiative recombination that remains to be determined. In equilibrium with the radiation in the environment, in which $n_e n_h = n_e^0 n_h^0 = n_i^2$,

$$G_\gamma^0 = R_e^0 = R_h^0 = R_\gamma^0 = G_e^0 = G_h^0 = Bn_e^0 n_h^0 \tag{3.56}$$

It follows immediately from this equation that the product $n_e^0 n_h^0$ cannot depend on the doping and must therefore be constant, with a value of n_i^2, as long as the absorption coefficient for band–band transitions and therefore the rate of absorption of photons R_γ^0 does not depend on the doping. Doping would have a significant effect on the absorption coefficient only if the density of impurities were in the same order of magnitude as the density of states in the bands.

The equality between G_γ^0 and R_γ^0, which, in equilibrium, applies not only in integral form over the spectrum but also in each photon energy interval $d\hbar\omega$, is the result of Kirchhoff's law of radiation. Microscopically, this means that in thermal and chemical equilibrium with the radiation of the environment the rates of all generation processes are compensated individually by equally large rates of recombination processes between the same initial and final states: this is the principle of detailed balance. For the calculation of G_γ^0 we therefore take the expression for G_e^0 from Equation 3.54

$$G_\gamma^0 = R_\gamma^0 = G_e^0 = G_h^0 = \frac{\Omega}{4\pi^3\hbar^3 c^2} \int_0^\infty \frac{\alpha(\hbar\omega)(\hbar\omega)^2}{\exp\left(\frac{\hbar\omega}{kT_0}\right) - 1} d\hbar\omega \tag{3.57}$$

Knowledge of the absorption coefficient $\alpha(\hbar\omega)$, determined from absorption measurements, and of n_i then allows the calculation of the coefficient B for radiative recombination in Equation 3.55.

The rate per volume at which photons in silicon in equilibrium with the 300 K black-body radiation of the environment are generated is given by

$$G_\gamma^0(\text{Si}) = 3 \times 10^5 \frac{\gamma}{\text{cm}^3\text{s}} = Bn_i^2$$

For silicon $n_i = 10^{10}\ \text{cm}^{-3}$ and therefore

$$B(\text{Si}) = 3 \times 10^{-15} \frac{\gamma\ \text{cm}^3}{\text{s}}$$

In a state of nonequilibrium between the electrons and holes in the semiconductor on one side and the photons in the 300 K radiation of the environment on the other side, resulting, e.g. from incident solar radiation or injection or extraction of electrons and holes with an electric current, $n_e n_h \neq n_i^2$. For extraction without additional production, in fact $n_e n_h < n_i^2$. The distribution of electrons and holes over the states is, however, still determined by Fermi distributions at the lattice temperature, with a mean kinetic energy of $\langle \varepsilon_{kin} \rangle = (3/2)kT_0$. In a state of nonequilibrium with the 300 K radiation of the environment, photons are consequently emitted with the same energy distribution as in equilibrium with the 300 K radiation, but at the different rate

$$G_\gamma = G_\gamma^0 \frac{n_e n_h}{n_i^2} \tag{3.58}$$

With reference to the quasi-Fermi energies in Equation 3.29 we can now write

$$G_\gamma = G_\gamma^0 \exp\left(\frac{\varepsilon_{FC} - \varepsilon_{FV}}{kT}\right) \tag{3.59}$$

Owing to solar irradiation, the emission rate of photons increases in proportion to $n_e n_h$. Since the annihilated electron–hole pairs are lost for the electric current that a solar cell shall deliver, it is of interest to know the total current of photons emitted by a semiconductor of thickness d. This is not just the emission rate in Equation 3.58 multiplied by the thickness of the semiconductor, since some of the emitted photons are reabsorbed and generate electron–hole pairs again.

Although this would be a good exercise in the application of the continuity equation, the integration of the emission and absorption rates of the photons over the volume of the semiconductor and the consideration of total internal reflection at the surface is a rather confusing calculation. Instead, we will once again make use of the principle of detailed balance to determine the total loss of electron–hole pairs in a cell due to radiative recombination in a more elegant way.

In thermal and chemical equilibrium with the radiation of the environment, the photon current density $dj_\gamma^0(\hbar\omega)$ emitted from the surface of a body into the environment must be equal to the photon current density incident from the environment and absorbed by the body:

$$dj_{\gamma,\text{em}}^0 = dj_{\gamma,\text{abs}}^0 = a(\hbar\omega)dj_\gamma^0$$

In nonequilibrium of homogeneous electron and hole distributions, the emitted photon current density is larger by the same factor $n_e n_h / n_i^2$ by which the emission rate in Equation 3.58 is increased, so that

$$dj_{\gamma,\text{em}}(\hbar\omega) = a(\hbar\omega)\frac{n_e n_h}{n_i^2} dj_\gamma^0(\hbar\omega) \tag{3.60}$$

Here $a(\hbar\omega)$ is the absorptance, which, without considering multiple internal reflection, is $a(\hbar\omega) = [1 - r(\hbar\omega)]\{1 - \exp[-\alpha(\hbar\omega)d]\}$. The absorptance is always $\leq 1 - r$ and for a thickness $d \gg 1/\alpha$, $a = 1 - r$, independent of the thickness. This means that the emitted photon current density reaches a limit value for large thickness, which then no longer depends on the thickness. Although the rate of radiative recombination integrated over the volume increases linearly with the volume, for a large thickness, most of the emitted photons are reabsorbed and generate electron–hole pairs again. The loss of electron–hole pairs by radiative recombination is given only by the photons that are emitted through the surface of the semiconductor. The effective rate of radiative recombination is proportional to the surface area and does not depend on the volume, if the thickness is larger than the penetration depth of the photons.

Expressing Equation 3.60 in terms of the quasi-Fermi energies, we obtain

$$dj_{\gamma,\mathrm{em}} = a(\hbar\omega)\exp\left[\frac{(\varepsilon_{FC}-\varepsilon_{FV})}{kT}\right]\frac{\Omega}{4\pi^3\hbar^3c^2}\frac{(\hbar\omega)^2}{\exp(\hbar\omega/kT)-1}\,d\hbar\omega \qquad (3.61)$$

This result is based on the Boltzmann approximation of the Fermi distribution in deriving the relationship between the concentrations of electrons and holes and the Fermi energies in Equations 3.12 and 3.15. For this approximation, the Fermi energies had to be a few kT away from the band edges. The same restriction applies to the absorption coefficient in the radiative recombination rate (per volume) and to the absorptance in the emitted photon current (per surface area). The condition of a full valence band and an empty conduction band on which their derivation in Section 3.5 is based and for which the absorption coefficient is usually measured is no longer a good approximation if the Fermi energies come close to the band edges. Equation 3.61 is therefore only approximately true. An exact derivation is given in Section 3.7.

In place of Equation 3.61 the exact result is [2]

$$dj_{\gamma,\mathrm{em}} = a(\hbar\omega)\frac{\Omega}{4\pi^3\hbar^3c^2}\frac{(\hbar\omega)^2}{\exp\left\{\dfrac{\hbar\omega-(\varepsilon_{FC}-\varepsilon_{FV})}{kT}\right\}-1}\,d\hbar\omega \qquad (3.62)$$

In this form, as the generalized Planck radiation law, it describes both the emission of thermal radiation for $(\varepsilon_{FC}-\varepsilon_{FV}) = 0$ and the emission of luminescence radiation for $(\varepsilon_{FC}-\varepsilon_{FV}) \neq 0$. The difference of the Fermi energies $(\varepsilon_{FC}-\varepsilon_{FV})$ turns out to be the chemical potential μ_γ of the emitted photons. For the operation of solar cells, where $(\varepsilon_{FC}-\varepsilon_{FV})$ is several kT smaller than ε_G, Equation 3.61 is, however, a very good approximation.

3.6.2 Nonradiative Recombination

With the recombination of an electron and a hole, the energy set free must always be taken up by other particles. For nonradiative recombination these are primarily other electrons or holes (Auger recombination) or phonons (impurity recombination).

Auger Recombination

Auger recombination is the reverse of impact ionization, in which an electron or a hole with high kinetic energy knocks another electron out of its bond, thereby creating a free electron and a free hole. In the reverse process, the energy set free during recombination is transferred to an electron or a hole as kinetic energy that is subsequently lost to the lattice through collisions with phonons. Figure 3.20

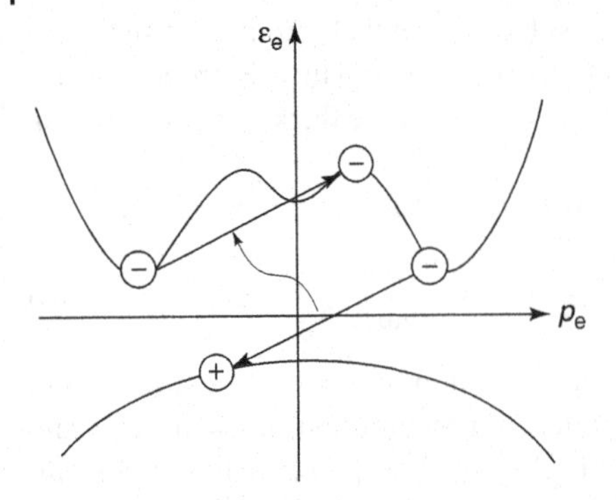

Figure 3.20 In Auger recombination, the energy set free by recombination of an electron–hole pair is absorbed by a free carrier (an electron in this figure) and is subsequently dissipated by generating phonons in collisions with the lattice.

illustrates the process of Auger recombination in which energy and momentum must initially be conserved.

If an electron absorbs the energy, two electrons and one hole participate in the reaction. The recombination rate is then

$$R_{\text{Aug, e}} = C_e n_e^2 n_h$$

and is large for strong n-doping.

If a hole absorbs the recombination energy, the recombination rate is

$$R_{\text{Aug, h}} = C_h n_e n_h^2$$

which is large in materials with strong p-doping. In general, the rate of Auger recombination is

$$R_{\text{Aug}} = n_e n_h (C_e n_e + C_h n_h) \tag{3.63}$$

In silicon the constants have the value

$$C_e(\text{Si}) \approx C_h(\text{Si}) \approx 1 \times 10^{-30}\ \text{cm}^6\ \text{s}^{-1}$$

A pn solar cell requires doping. How strong the doping should be will be discussed later. For this structure, Auger recombination represents a practically unavoidable loss and largely determines the efficiency limits of the best silicon solar cells.

Recombination via Impurities

In equilibrium with 300 K radiation of the environment, for which $n_e^0 n_h^0 = n_i^2$ and the generation of electrons by photons is compensated by radiative recombination, nonradiative recombination processes are also compensated by the reverse processes, i.e. nonradiative generation processes in which electrons are excited via states in the energy gap or, less frequently, as a result of the absorption of many phonons, even directly from the valence band to the conduction band.

In real solar cells, recombination via impurities is the predominant recombination process and will therefore be discussed in detail. Impurities that offer electron states with energies approximately in the middle of the forbidden zone play the most important role. They capture electrons and holes over a series of excited states, with successive dissipation of energy. Since the recombination energy can thus be imparted to the lattice in small portions through the production of individual phonons, the nonradiative transition of an electron from the conduction band to the valence band, which is tantamount to the capture of an electron and a hole by an impurity, can take place much more easily. In the following analysis, we will therefore neglect radiative recombination compared with nonradiative recombination. The rates of generation and recombination involving impurities and considered in the analysis, are shown in Figure 3.21. The analysis follows the calculations of Shockley, Read, and Hall [3].

As an example, we will examine a single type of acceptor-like impurity, having a concentration n_{imp} with a known electron energy ε_{imp} and a known capture cross section σ_e for electrons and σ_h for holes in a homogeneously excited semiconductor under open circuit conditions, i.e. without current flow. Under these conditions all electrons and holes must eventually recombine. During their lifetime they will distribute homogeneously by diffusion. This allows to replace the actual and inhomogeneous generation rate of electron–hole pairs due to illumination by an average over the total volume of a semiconductor. Since for this consideration recombination via impurities is assumed to be dominant, no other recombination mechanism is considered.

The electron concentration n_e (in the conduction band) changes in time owing to the rate of generation $G_e = G_h = G = a(\hbar\omega) j_\gamma / d$ by the absorption of photons averaged over the thickness d and it changes by the rate $R_{e,imp}$ of capture into impurities occupied by holes and by the rate of thermal emission $G_{e,imp}$ from impurities occupied by electrons. The contributions of the 300 K radiation from the environment and radiative recombination are neglected. The continuity equations for the changes in the concentrations of electrons (in the conduction band) n_e, of holes (in the valence band) n_h and of occupied impurity states, i.e. of electrons

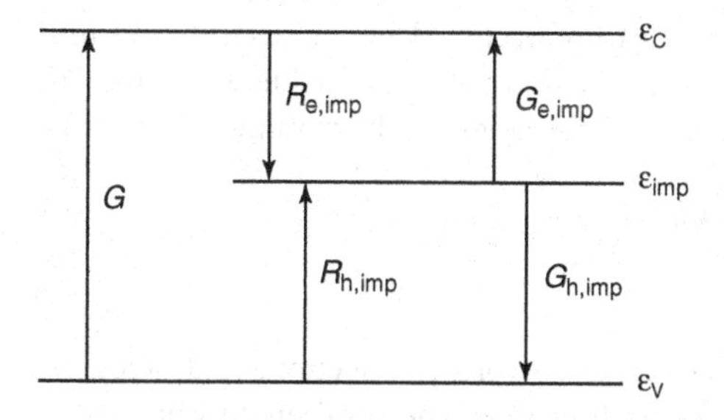

Figure 3.21 Electrons and holes produced by the absorption of photons with the generation rate G are captured by impurities with the rates $R_{e,imp}$ and $R_{h,imp}$ and they are emitted thermally with the rates $G_{e,imp}$ and $G_{h,imp}$ from the impurities back into the bands.

trapped in impurities $n_{e,imp}$ as a function of the generation and recombination rates shown in Figure 3.21 are

$$\frac{\partial n_e}{\partial t} = G - R_{e,imp} + G_{e,imp} \tag{3.64}$$

$$\frac{\partial n_h}{\partial t} = G - R_{h,imp} + G_{h,imp} \tag{3.65}$$

$$\frac{\partial n_{e,imp}}{\partial t} = R_{e,imp} - R_{h,imp} - G_{e,imp} + G_{h,imp} \tag{3.66}$$

These are three equations for the three unknown concentrations n_e, n_h and $n_{e,imp}$. Unfortunately, these equations are not linearly independent, so that, e.g. the third equation follows from the difference of the first and the second equation. We obtain the missing third independent equation from the condition that charge neutrality of the semiconductor must prevail even when the different concentrations are changing.

$$e(n_{D^+} - n_e - n_A^- + n_h - n_{e,imp}) = 0 \tag{3.67}$$

In this equation we assume that the impurities are negatively charged when they are occupied by electrons, i.e. they are acceptor-type impurities. For donor-type impurities, positively charged when not occupied by electrons, $(-n_{e,imp})$ must be replaced by $(n_{imp} - n_{e,imp})$, where n_{imp} is the concentration of the impurities.

We now have to determine the various generation and recombination rates. The rate of recombination $R_{e,imp}$ at which electrons vanish from the conduction band follows from a purely kinetic argument. Whenever an electron that moves through the lattice at its thermal velocity v_e comes within a capture cross section σ_e of an impurity that is occupied by a hole, it will be captured. Impurities that are already occupied by an electron will not capture a second electron, because the energy of a second electron state is much higher and may lie in the energy range of the conduction band. The rate at which capture happens to all electrons is

$$R_{e,imp} = \sigma_e v_e n_e n_{h,imp} \tag{3.68}$$

where $n_{h,imp}$ is the density of impurities not occupied by an electron and thus occupied by a hole, which are the recombination partners for the electrons. The capture cross sections for electrons, σ_e, and for holes, σ_h, have values of the order of 10^{-15} cm^2.

Similarly, the recombination rate for holes is given by

$$R_{h,imp} = \sigma_h v_h n_h n_{e,imp} \tag{3.69}$$

An impurity is likely to have different cross sections for electrons and holes. An unoccupied donor-type impurity is positively charged and consequently has a large capture cross section for electrons, while an unoccupied acceptor-type impurity is neutral, with a small capture cross section for electrons. Conversely, the capture cross section for holes is large for an occupied, and therefore negative, acceptor impurity and is small for an occupied and neutral donor impurity.

In Equations 3.68 and 3.69 the occupation of impurity states, and as a result the concentration of electrons $n_{e,imp}$ or holes $n_{h,imp}$ in the impurities, is not known. If the energy ε_{imp} of the electrons in the impurities lies between the quasi-Fermi energies, the occupation of the impurities is not defined by either of the Fermi distributions but is rather determined by the kinetics.

For the generation rates of electrons or holes from the impurities into the bands, we know that they must be proportional to the concentration of electrons $n_{e,imp}$ or holes $n_{h,imp} = n_{imp} - n_{e,imp}$ in the impurities. The generation rate of electrons is

$$G_{e,imp} = \beta_e n_{e,imp} \tag{3.70}$$

In the same way, the generation rate of holes is

$$G_{h,imp} = \beta_h n_{h,imp} = \beta_h (n_{imp} - n_{e,imp}) \tag{3.71}$$

Here β_e and β_h are as yet unknown coefficients of emission. For their determination we make use of the principle of detailed balance in the dark, inherent in the thermal and chemical equilibrium with radiation from the environment. According to Equations 3.64, 3.68, and 3.70, in the absence of external excitation ($G = 0$) and in a steady state

$$\frac{\partial n_e}{\partial t} = -\sigma_e v_e n_e (n_{imp} - n_{e,imp}) + \beta_e n_{e,imp} = 0 \tag{3.72}$$

From Equation 3.72 we can calculate the density of impurities occupied by electrons in the dark, which we already know is given by the equilibrium Fermi distribution,

$$n_{e,imp} = n_{imp} \frac{1}{\beta_e/(\sigma_e v_e n_e) + 1} = n_{imp} \frac{1}{\exp\left[(\varepsilon_{imp} - \varepsilon_F)/kT\right] + 1} \tag{3.73}$$

With $n_e = N_C \exp[-(\varepsilon_C - \varepsilon_F)/kT]$, we find for the coefficient of emission of electrons from the impurity into the conduction band

$$\beta_e = \sigma_e v_e N_C \exp\left(-\frac{\varepsilon_C - \varepsilon_{imp}}{kT}\right) \tag{3.74}$$

and similarly for the coefficient of emission of holes from the impurity into the valence band

$$\beta_h = \sigma_h v_h N_V \exp\left(-\frac{\varepsilon_{imp} - \varepsilon_V}{kT}\right) \tag{3.75}$$

The capture cross sections and the coefficients of emission are likely to depend on the energy distribution of the free charge carriers. Owing to the rapid thermalization, however, electrons and holes have the same energy and velocity distributions under illumination as in the dark. Capture cross sections and emission coefficients are therefore expected to have the same values both in the dark and under illumination.

We now have all the ingredients to solve Equations 3.64–3.67, which define the concentrations of the electrons, holes, and occupied (as well as unoccupied) impurities as a function of the external rate of generation G of electrons and holes, and not only in the steady state. For non-steady-state processes this system of equations can, however, only be solved numerically.

For the steady-state case, we find from Equation 3.66 for the density of impurities occupied by electrons

$$n_{e,\,imp} = \frac{n_{imp}\left\{\sigma_e v_e n_e + \sigma_h v_h N_V \exp\left[-(\varepsilon_{imp}-\varepsilon_V)/kT\right]\right\}}{\sigma_e v_e\left\{n_e + N_C \exp\left[-(\varepsilon_C-\varepsilon_{imp})/kT\right]\right\} + \sigma_h v_h\left\{n_h + N_V \exp\left[-(\varepsilon_{imp}-\varepsilon_V)/kT\right]\right\}} \tag{3.76}$$

This result, which is not important by itself, inserted into Equation 3.64 gives

$$G = \frac{n_e n_h - n_i^2}{\dfrac{n_e + N_C \exp\left[-(\varepsilon_C-\varepsilon_{imp})/kT\right]}{n_{imp}\sigma_h v_h} + \dfrac{n_h + N_V \exp\left[-(\varepsilon_{imp}-\varepsilon_V)/kT\right]}{n_{imp}\sigma_e v_e}} \tag{3.77}$$

The left side represents the rate of generation of both electrons and holes. For the steady-state case, then, on the right side we have the recombination rate, with the same value for both electrons and holes. Here $n_{imp}\sigma_h v_h$ is the rate per hole of hole capture if all impurities are occupied by electrons. Its reciprocal value $\tau_{h,\,min} = 1/(n_{imp}\sigma_h v_h)$ is the smallest mean lifetime of a hole in the valence band between generation and capture, since it implies that *all* impurities are occupied by electrons. Similarly, $\tau_{e,\,min} = 1/(n_{imp}\sigma_e v_e)$ is the mean lifetime of an electron in the conduction band when *all* impurities are occupied by holes, which is the smallest possible value of the electron lifetime. The minimal lifetimes defined in this way represent lower limits for the actual lifetimes when the impurities are only partially occupied.

Since separation of the Fermi energies is the main objective of a solar cell, it is worthwhile to express Equation 3.77 in terms of the Fermi energies. We use the relations

$$n_e = n_i \exp\left(-\frac{\varepsilon_i - \varepsilon_{FC}}{kT}\right) \tag{3.78}$$

$$n_h = n_i \exp\left(-\frac{\varepsilon_{FV} - \varepsilon_i}{kT}\right) \tag{3.79}$$

$$N_C \exp\left(-\frac{\varepsilon_C - \varepsilon_{imp}}{kT}\right) = n_i \exp\left(-\frac{\varepsilon_i - \varepsilon_{imp}}{kT}\right) \tag{3.80}$$

$$N_V \exp\left(-\frac{\varepsilon_{imp} - \varepsilon_V}{kT}\right) = n_i \exp\left(-\frac{\varepsilon_{imp} - \varepsilon_i}{kT}\right) \tag{3.81}$$

where ε_i is the value of the Fermi energy in an intrinsic semiconductor in the dark, and we find from Equation 3.77

$$G = \frac{n_i\left\{\exp\left[(\varepsilon_{FC}-\varepsilon_{FV})/kT\right]-1\right\}}{\tau_{h,\,min}\left\{\exp\left[(\varepsilon_{FC}-\varepsilon_i)/kT\right]+\exp\left[(\varepsilon_{imp}-\varepsilon_i)/kT\right]\right\} + \tau_{e,\,min}\left\{\exp\left[(\varepsilon_i-\varepsilon_{FV})/kT\right]+\exp\left[(\varepsilon_i-\varepsilon_{imp})/kT\right]\right\}} \tag{3.82}$$

For a good solar cell we want the difference in the Fermi energies, i.e. the numerator of Equation 3.82, to be large. For a given generation rate G, the denominator must then be large, too, requiring large minimal lifetimes. Conversely, a material produces a poor solar cell when the numerator and the denominator of Equation 3.82 are both small. If the minimum lifetimes of the holes $\tau_{h,min}$ and electrons $\tau_{e,min}$ are equal, the denominator will be minimal if, on the one hand, the impurity level ε_{imp} coincides with the position of the Fermi energy in the intrinsic state ($\varepsilon_{imp} = \varepsilon_i$) and, on the other hand, if the Fermi energies ε_{FC} and ε_{FV} are symmetrical about the intrinsic Fermi energy ε_i, which is approximately in the middle of the energy gap ($\varepsilon_{FC} - \varepsilon_i = \varepsilon_i - \varepsilon_{FV}$). In an homogeneous material the latter condition is fulfilled only if it is intrinsic. Doping, regardless of whether n-type or p-type, will increase the difference in the Fermi energies for a given generation rate compared with an intrinsic material.

For equal minimal lifetimes of electrons and holes the difference in the Fermi energies $\varepsilon_{FC} - \varepsilon_{FV}$ for a constant generation rate G is determined from Equation 3.82 and is shown in Figure 3.22 as a function of the impurity level ε_{imp}. As expected, $\varepsilon_{FC} - \varepsilon_{FV}$ is small when the electrons in the impurities have energies ε_{imp} in the middle of the energy gap. The damaging influence of the impurities is also greater in weakly doped material, where the Fermi energies are more nearly symmetrical about the middle of the energy gap than in strongly doped material. Whether the material is p-doped or n-doped, however, is irrelevant. The type and level of doping can be seen from the position of the Fermi energy ε_F^0 in the dark state. The greater the separation $|\varepsilon_F^0 - \varepsilon_i|$ of the equilibrium Fermi energy from the intrinsic Fermi

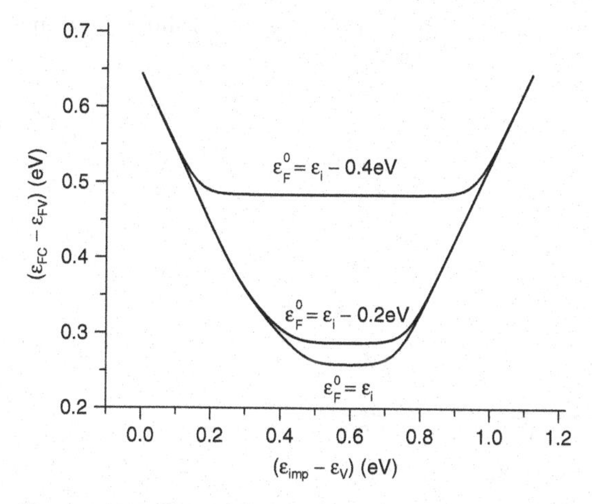

Figure 3.22 Difference in the Fermi energies as a function of the impurity energy ε_{imp} for a constant rate of generation $G = 10^{17}$ cm^{-3}s^{-1} and for different doping levels in a material with $\varepsilon_G = 1.12$ eV. The doping concentration can be inferred from the difference between the Fermi energy in the dark ε_F^0 and the intrinsic Fermi energy ε_i from Equation 3.78 or 3.79. The impurities have a concentration of $n_{imp} = 10^{14}$ cm^{-3} with equal capture cross sections for electrons and holes of $\sigma = 10^{-15}$ cm^2, the velocity of electrons and holes is $v = 10^5$ m s^{-1}, resulting in a minimal lifetime of $\tau = 10^{-6}$ s. The asymmetry is caused by the acceptor character of the impurities.

energy, i.e. the closer the equilibrium Fermi energy is to one of the band edges, the stronger the doping of the material. For n-doped material $\varepsilon_F^0 - \varepsilon_i > 0$ and for p-doped material $\varepsilon_F^0 - \varepsilon_i < 0$.

The concentrations of additional electrons and holes due to the external generation rate G can be determined analytically from Equation 3.77 for small impurity concentrations n_{imp} and weak excitation. Furthermore, we will divide the charge carrier densities into those in the dark state and those produced by the external generation, $n_e = n_e^0 + \Delta n_e$ and $n_h = n_h^0 + \Delta n_h$, respectively. From the charge neutrality condition in Equation 3.67 it follows that for a small impurity concentration $n_{imp} \ll \Delta n_e, \Delta n_h$ the additional concentrations of electrons and holes are equal, $\Delta n_e = \Delta n_h = \Delta n$. The condition of weak excitation is fulfilled, if $\Delta n \ll n_e^0 + n_h^0$. With these restrictions, it follows from Equation 3.77 for capture cross sections not too different for electrons and holes that

$$\Delta n = G \left\{ \tau_{h,\min} \frac{n_e^0 + N_C \exp\left[-(\varepsilon_C - \varepsilon_{imp})/kT\right]}{n_e^0 + n_h^0} + \tau_{e,\min} \frac{n_h^0 + N_V \exp\left[-(\varepsilon_{imp} - \varepsilon_V)/kT\right]}{n_e^0 + n_h^0} \right\} \tag{3.83}$$

Figure 3.23 shows the ratio $\Delta n/G$, which is also the mean lifetime τ of electrons and holes, as a function of ε_{imp} in p-doped silicon. $\tau_{e,\min} = \tau_{h,\min} = 10^{-6}$ s was used for the calculation, resulting from an impurity density $n_{imp} = 10^{14}$ cm^{-3}, a capture cross section for electrons and holes of $\sigma = 10^{-15}$ cm^2 and a velocity of $v = 10^5$ m s^{-1}. We see in Figure 3.23 that the lifetime τ becomes shorter and the

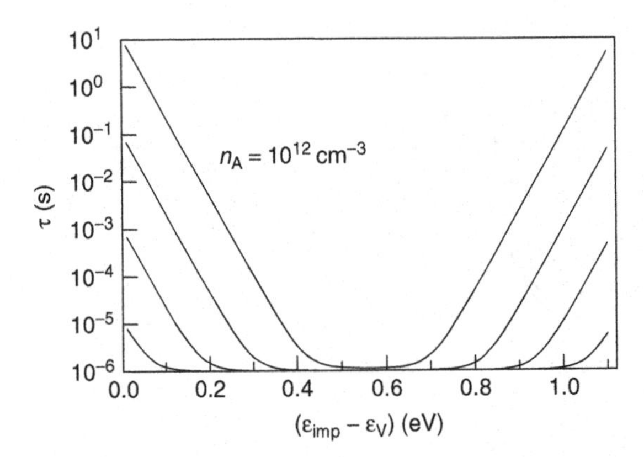

Figure 3.23 Lifetime τ of electrons and holes in p-doped silicon limited by impurity recombination as a function of the impurity energy ε_{imp}, measured from the upper edge of the valence band. With each curve from the inside toward the outside, the acceptor density increases by a factor of 100 from $n_A = 10^{12}$ cm^{-3} to $n_A = 10^{18}$ cm^{-3}. The generation rate is $G = 10^{17}$ cm^{-3} s^{-1}, the impurities have a concentration of $n_{imp} = 10^{14}$ cm^{-3} with equal capture cross sections for electrons and holes of $\sigma = 10^{-15}$ cm^2 and the velocity of electrons and holes is $v = 10^5$ m s^{-1}, resulting in a minimal lifetime of $\tau_{min} = 10^{-6}$ s.

recombination accordingly more effective as the energy level of the impurity ε_{imp} approaches the intrinsic Fermi energy, which is in the middle of the forbidden zone, if the effective masses of electrons and holes are equal.

When the impurity level is close to the conduction band, the impurities are less effective, because most of the time they are not occupied by electrons. Captured electrons are much more frequently emitted back to the conduction band than annihilated by recombination with a hole captured from the valence band. When the impurity level is in the vicinity of the valence band, impurities are most of the time occupied by electrons, and captured holes are emitted back to the valence band before they can recombine by electron capture. In the middle of the forbidden zone, occupation by electrons and holes is equally probable. Both electrons and holes find enough free impurity states to be captured and reemission is less probable. We also recognize the extreme demands on the concentration of impurities with energies in the middle of the gap and therefore on the purity of the material in order to obtain lifetimes of a few milliseconds, as is achieved in good silicon solar cells.

A large difference of the Fermi energies and a large lifetime of electrons and holes are both indications of a good-quality solar cell material. A closer look at Figures 3.22 and 3.23 seems to reveal an inconsistency. While increasing doping concentrations lead to an increasing difference of the Fermi energies in Figure 3.22, they result in decreasing lifetimes in Figure 3.23. The explanation is simple. In a p-type semiconductor under low excitation ($\Delta n \ll n_h^0$) the difference of the Fermi energies is $\Delta\varepsilon_F = kT \ln\left(n_e n_h^0/n_i^2\right)$. As long as the lifetime τ and with it the electron concentration $n_e = \Delta n = G\tau$ decreases less strongly than the doping concentration and with it n_h^0 increases, the difference of the Fermi energies will rise with the doping concentration, although the lifetime decreases. The question of whether the splitting of the Fermi energies or the lifetime is the more important quantity for a solar cell will be addressed later.

If, as in a real material, impurity states with different energies ε_{imp} and different capture cross sections are present, then contributions from each of them must be added to the recombination rate R_{imp}, the generation rate G_{imp}, and the charge neutrality condition in Equations 3.64–3.67. But as we have seen in Figures 3.22 and 3.23, impurity states in the middle of the band gap dominate the recombination properties.

Surface Recombination

States with energies in the forbidden gap, which are very effective mediators for nonradiative recombination are found in large concentrations on the surface of a material where neighboring atoms are no longer available to the lattice atoms and where impurity molecules (e.g. O_2 or H_2O) are adsorbed. These so-called surface states often have a continuous distribution over energy in the forbidden gap, as illustrated in Figure 3.24. The effect of the surface states on nonradiative recombination can be treated in the same way as the effect of the bulk states before. If the concentration of the surface states per energy and their cross section for capture of electrons and holes are known as a function of energy,

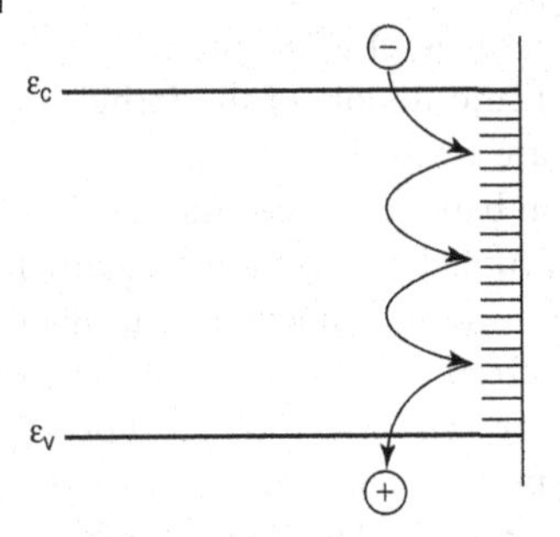

Figure 3.24 Recombination via surface states continuously distributed over energy in the energy gap of a semiconductor.

the total recombination rate is the sum over the contributions from each energy interval.

In a simpler treatment, all the contributions are lumped into a single expression. For electrons, for example, the recombination rate via surface states is

$$R_{s,e} = \sigma_{s,e}\, v_e n_{s,h} n_e \tag{3.84}$$

Here $R_{s,e}$ is the recombination rate of electrons per area, $n_{s,h}$ is the density of surface states per area occupied by a hole, and n_e is the concentration of electrons at the surface. The product $\sigma_{s,e} v_e n_{s,h}$ has the dimension of velocity and is known as the *surface recombination velocity* $v_{R,e}$ of the electrons. It is characteristic for the surface quality, but it is not a material property alone, since it depends on the occupation of the surface states and thus on the excitation conditions,

$$R_{s,e} = v_{R,e} n_e \tag{3.85}$$

As with impurities in the bulk of a semiconductor, surface states are most effective with respect to recombination when their electron energies lie in the middle of the forbidden gap and in between the Fermi energies. For this reason nonradiative recombination (at the surface) is reduced, if both Fermi energies are close to either the conduction band or the valence band at the surface, as is the case for heavy doping. Much less is known about the chemical nature of the surface states and their electron energies compared to impurities in the bulk of a semiconductor. In most cases, we must restrict their characterization to the surface recombination velocity.

Poor surfaces with high surface recombination velocities of 10^5–10^6 cm s^{-1} are those freely exposed to air, which adsorb H_2O and O_2 or react chemically with these substances.

Metallic surfaces required as contacts for the supply of electric current are especially problematic. As can be seen in Figure 3.25, the forbidden gap of the semiconductor is adjacent to the continuous distribution of states in the conduction band of the metal. To a good approximation, we can consider the surface recombination velocity at a metallic contact to be $v_{R,\,metal} \approx \infty$. The resulting infinitely large recombination rate is compensated in the dark by a generation rate that is

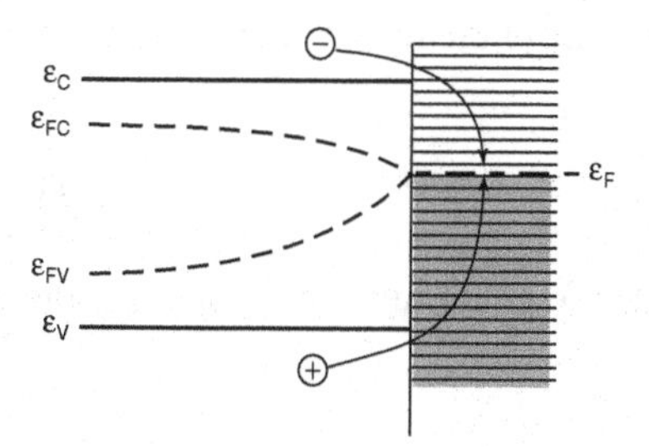

Figure 3.25 The continuous distribution of a high density of states around the Fermi energy in a metal results in a high interface recombination velocity at a semiconductor–metal contact.

likewise infinitely large. Any additional generation rate at the semiconductor/metal interface due to light absorption is negligibly small compared with the dark rate. As a result, even with additional generation, the concentrations of electrons and holes at a metallic contact do not differ from their equilibrium values in the dark state for which $n_e^0 n_h^0 = n_i^2$. In the same way, the Fermi energies are not separated at a metal contact.

With silicon, good surfaces can be prepared by carefully controlled oxidation under clean room conditions. Surface state densities of $D_s < 10^{10}\ (\mathrm{cm}^2\ \mathrm{eV})^{-1}$ can be obtained at Si/SiO_2 interfaces, with recombination velocities of $v_R \leq 10\ \mathrm{cm\ s}^{-1}$.

Such perfect surface passivation is possible only for Si/SiO_2 and for the combination of a few III–V compounds having the same lattice constants. In spite of this, contact with another semiconductor in the form of a cover layer usually leads to lower surface recombination velocities than for a free surface or for the contact with a metal. In order for the recombination on the surface of the cover layer to remain insignificant, no electron–hole pairs should be generated within this layer. Semiconductors with a large band gap ($\varepsilon_G > 3$ eV) are therefore chosen for passivating cover layers. These are transparent in the visible range and are also known as *window layers*.

3.6.3 Lifetimes

Because of recombination, a charge carrier once produced does not "survive" arbitrarily long in its band. If electrons and holes are produced by the absorption of light and the light source is suddenly switched off, the charge carriers then vanish after a mean lifetime τ. Once again, the continuity equation tells us how:

$$\frac{\partial n_e}{\partial t} = G_e - R_e - \operatorname{div} j_e$$

We will restrict ourselves here to a spatially homogeneous system ($\operatorname{div} j_e = 0$) and divide the concentrations and rates into what is present in equilibrium, with $n_e^0 n_h^0 = n_i^2$, and what is caused by deviations from equilibrium. This division of the recombination rate, e.g. for electrons, is meaningful only if it changes in proportion to the electron density n_e, that is, when the density of the recombination partners, the holes, is constant:

$$\frac{\partial(n_e^0 + \Delta n_e)}{\partial t} = G_e^0 + \Delta G_e - (R_e^0 + \Delta R_e) \tag{3.86}$$

Since $\partial n_e^0/\partial t = G_e^0 - R_e^0 = 0$ characterizes the equilibrium dark state, for the additional concentration of electrons

$$\frac{\partial \Delta n_e}{\partial t} = \Delta G_e - \Delta R_e$$

On suddenly switching off ΔG_e, we have for radiative recombination

$$\frac{\partial \Delta n_e}{\partial t} = -\Delta R_e = -B n_h \Delta n_e \tag{3.87}$$

This equation assumes that the generation of electrons and holes can be switched off, because it is caused entirely by photons incident from outside. It does not consider the reabsorption of the photons emitted by radiative recombination. The equation is therefore valid only for small semiconductor thicknesses. If we also consider the reabsorption of photons produced in thick bodies, the electron density falls off more slowly by radiative recombination as the thickness of the body increases. The effective rate of radiative recombination is only caused by the photons leaving the semiconductor through the surface and is thus surface related in contrast to nonradiative recombination, which is volume related. For semiconductors thicker than the penetration depth of the photons the effective rate of radiative recombination saturates and becomes independent of the thickness. From then on the radiative lifetime increases with the thickness.

Equation 3.87 can be integrated easily for the electrons in a heavily doped p-conductor in which, because of the strong p-doping, the density of holes $n_h = n_h^0 + \Delta n_h \approx n_h^0$ is nearly independent of illumination. We refer to this case, in which the additional concentrations are small relative to the majority carrier concentration in the dark state, as "weak excitation" or "weak injection". For solar cells in unfocused solar radiation, this condition is fulfilled. Equation 3.87 then yields

$$\Delta n_e(t) = \Delta n_e(0) \exp(-t/\tau_{e,\mathrm{rad}}) \tag{3.88}$$

The characteristic time $\tau_{e,\mathrm{rad}} = 1/(B\, n_h^0)$ is called the *lifetime* (of the minority charge carriers, here electrons) for radiative recombination.

In p-type silicon, with a typical hole density of $n_h^0 = 10^{16}\ \mathrm{cm}^{-3}$, the lifetime for electrons with respect to radiative recombination is

$$\tau_{e,\mathrm{rad}} = \frac{1}{3 \times 10^{-15}\ \mathrm{cm}^3\ \mathrm{s}^{-1} \times 10^{16}\ \mathrm{cm}^{-3}} = 0.03\ \mathrm{s}$$

Making use of the lifetime, we can write the recombination rate in general as

$$\Delta R_e = \frac{\Delta n_e}{\tau_e}$$

Suddenly switching on ΔG_e starting from a state of equilibrium, we obtain

$$\Delta n_e(t) = \Delta G_e \tau_e \left\{1 - \exp\left(-\frac{t}{\tau_e}\right)\right\} \tag{3.89}$$

The additional steady-state concentration of electrons resulting from an additional generation rate by light absorption is

$$\Delta n_e = \Delta G_e \tau_e \tag{3.90}$$

For impurity recombination in a steady state, once again $\Delta R_e = \Delta G_e = \Delta n_e/\tau_e$ and thus

$$\tau_e = \frac{1}{\sigma_e v_e n_{h,\,imp}} \tag{3.91}$$

Figure 3.23 shows this lifetime as a function of the position of the impurity level.

If the density of the recombination partners varies with time, the lifetime also changes with time. We then speak of the momentary lifetime. Everything said here for the electrons in a p-conductor applies equally for the holes in a n-conductor and in general for the respective minority charge carriers.

The density of the free charge carriers can be monitored by the microwave absorption/reflection they cause. From the decay following the switching off of the illumination, the lifetime is experimentally determined, which is an indicator of the quality of the semiconductor material.

As a result of transitions between the valence band and the conduction band, electrons and holes are created in pairs, so that $\Delta G_e = \Delta G_h$. If only donors or acceptors are present, completely ionized at room temperature, and otherwise there are only negligibly few impurities, the conservation of charge in Equation 3.67 then requires that $\Delta n_e = \Delta n_h$ and therefore $\tau_e = \tau_h$ as well. This is a surprising result in view of the large differences in the densities of electrons and holes in a doped semiconductor.

The different recombination processes take place in parallel to each other in a semiconductor. The total recombination rate is the sum of the rates for the different recombination mechanisms. If each of them is characterized by a lifetime τ_i, which the electrons or holes would have if no other recombination process were present, then the total lifetime follows from

$$R_{tot} = \frac{\Delta n_e}{\tau_{e,tot}} = \sum_i R_i = \sum_i \frac{\Delta n_e}{\tau_i}$$

to be

$$\frac{1}{\tau_{tot}} = \sum_i \frac{1}{\tau_i}$$

If no electron–hole pairs are extracted, i.e. at open circuit, all charge carriers produced must eventually recombine, and $\Delta G_e = \Delta R_e = \Delta n_e/\tau_{e,\,tot}$. The additional concentration of electrons Δn_e in steady state is proportional to their lifetime.

The upper limit for the lifetime of electrons is $\tau_{e,rad}$ if all recombination mechanisms are avoided, except for radiative recombination, which is unavoidable. Since the maximum obtainable concentrations of electrons and holes occur if there is only radiative recombination, the knowledge of radiative recombination and of the emitted photon currents is of great importance for the determination of maximum efficiencies.

3.7
Light Emission by Semiconductors

Light is emitted by radiative recombination of electron–hole pairs. This light emission can be used as a diagnostic tool for the quality of the semiconductor. The generalized Planck equation in (3.62) describes the emission of light by any material as a function of its absorptance and of the chemical potential of its electron–hole pairs and these quantities can be deduced from the emitted light. The generalized Planck relation is so important that a rigorous derivation shall confirm its validity.

3.7.1
Transition Rates and Absorption Coefficient

The generalized Planck law for the emission of photons is derived in the same way as Planck's original formula, except that the body absorbing or emitting radiation contains electron distributions in separate energy ranges, which require different Fermi functions. This has consequences for the absorption coefficient and the probability of photon emission.

We analyze the absorption and emission rates per volume of photons by a semiconductor for radiative transitions between states in energy ranges of width $d\varepsilon$ at energies ε_1 in the valence band and ε_2 in the conduction band involving photons with energies in the energy range $\hbar\omega \ldots \hbar\omega + d\hbar\omega$. The states are occupied according to two different Fermi functions $f(\varepsilon_1)$ and $f(\varepsilon_2)$. Other transitions between states with the same energy difference contributing to the interaction with photons of the same energy $\hbar\omega$ are disregarded for the moment.

The rate of upward transitions from states at ε_1 to states at ε_2 induced by an incident photon current density dj_γ is

$$dr_{up}(\hbar\omega) = |M|^2 D_{12} f(\varepsilon_1)\left[1 - f(\varepsilon_2)\right] \frac{dj_\gamma(\hbar\omega)}{d\hbar\omega} d\hbar\omega \tag{3.92}$$

where M contains the matrix element for the transition and D_{12} is the combined density of states between which the transitions occur.

The rate of stimulated emission of photons is likewise

$$dr_{stim}(\hbar\omega) = |M|^2 D_{12}\left[1 - f(\varepsilon_1)\right] f(\varepsilon_2) \frac{dj_\gamma(\hbar\omega)}{d\hbar\omega} d\hbar\omega \tag{3.93}$$

and the rate of spontaneous emission finally is

$$dr_{\mathrm{spont}}(\hbar\omega) = |M|^2 \frac{c_0}{n} D_\gamma(\hbar\omega) D_{12} \left[1 - f(\varepsilon_1)\right] f(\varepsilon_2)\, d\hbar\omega \tag{3.94}$$

where the density of states for photons in the solid angle Ω in a medium with refractive index n is

$$D_\gamma(\hbar\omega) = \frac{\Omega n^3}{4\pi^3 \hbar^3 c_0^3} (\hbar\omega)^2 \tag{3.95}$$

and c_0 is the vacuum velocity of light.

Stimulated emission is a process by which a photon is duplicated, producing one additional photon in exactly the same state as the incident photon initiating the transition. These photons, therefore, cannot be distinguished from nonabsorbed photons. The net absorption rate at which photons disappear and which defines the absorption coefficient $\alpha_{12}(\hbar\omega)$ then is

$$\begin{aligned} dr_{\mathrm{abs}}(\hbar\omega) &= dr_{\mathrm{up}}(\hbar\omega) - dr_{\mathrm{stim}}(\hbar\omega) \\ &= |M|^2 D_{12} \left[f(\varepsilon_1) - f(\varepsilon_2)\right] dj_\gamma(\hbar\omega) = \alpha_{12}(\hbar\omega)\, dj_\gamma(\hbar\omega) \end{aligned} \tag{3.96}$$

resulting in an absorption coefficient

$$\alpha_{12}(\hbar\omega) = |M|^2 D_{12} \left[f(\varepsilon_1) - f(\varepsilon_2)\right] \tag{3.97}$$

This relation is different from the derivation of the absorption coefficient in Section 3.5.1, where an occupied valence band ($f(\varepsilon_1) = 1$) and an empty conduction band ($f(\varepsilon_2) = 0$) had been assumed. The important relation (3.97) allows us to replace the factor M and the combined density of states D_{12} in the spontaneous emission rate by the absorption coefficient $\alpha_{12}(\hbar\omega)$ which makes this treatment applicable to real materials for which only the absorption coefficient is known:

$$dr_{\mathrm{spont}}(\hbar\omega) = \alpha_{12}(\hbar\omega) \frac{c_0}{n} D_\gamma(\hbar\omega) \frac{\left[1 - f(\varepsilon_1)\right] f(\varepsilon_2)}{f(\varepsilon_1) - f(\varepsilon_2)} d\hbar\omega \tag{3.98}$$

With the Fermi functions for the states at energy ε_1 in the valence band for which the Fermi energy is ε_{FV} and for ε_2 in the conduction band for which the Fermi energy is ε_{FC}:

$$f(\varepsilon_1) = \frac{1}{\exp\left[(\varepsilon_1 - \varepsilon_{FV})/kT\right] + 1} \quad \text{and} \quad f(\varepsilon_2) = \frac{1}{\exp\left[(\varepsilon_2 - \varepsilon_{FC})/kT\right] + 1}$$

and remembering that the energy difference $\varepsilon_2 - \varepsilon_1$ over which the transitions occur is equal to the photon energy $\hbar\omega$, Equation 3.98 becomes

$$dr_{\mathrm{spont}}(\hbar\omega) = \alpha(\hbar\omega) \frac{c_0}{n} D_\gamma(\hbar\omega) \frac{d\hbar\omega}{\exp\left\{[\hbar\omega - (\varepsilon_{FC} - \varepsilon_{FV})]/kT\right\} - 1} \tag{3.99}$$

Since Equation 3.99 does not depend on the energies ε_1 and ε_2 explicitly, but only on their difference $\hbar\omega$, each pair of states with energies ε_j in the conduction band and ε_i in the valence band with the same energy difference $\varepsilon_j - \varepsilon_i = \hbar\omega$ contributes to Equation 3.99 in the same way. All possible transitions between valence band and conduction band with the photon energy $\hbar\omega$ are accounted for and are contained in Equation 3.99, if $\alpha(\hbar\omega)$ now is the absorption coefficient for all these transitions.[1)]

A problem is that the emission rate of the photons cannot be observed. What can be observed and measured is the photon current emitted through a surface. To find the emitted photon current, we have to integrate the difference between photon emission and absorption rates over the thickness of the semiconductor

$$\operatorname{div} \mathrm{d}j_\gamma(x) = \mathrm{d}r_{\mathrm{spont}}(x) - \alpha\, \mathrm{d}j_\gamma(x) \tag{3.100}$$

We will treat only the simple case of a homogeneously excited semiconductor, in which $\mathrm{d}r_{\mathrm{spont}}$ is constant. Far from the surface in an infinitely thick semiconductor, emission and absorption rates balance each other and the photon current is constant for $x \to \infty$. This equilibrium photon current inside the semiconductor is, from Equation 3.100,

$$\mathrm{d}j_\gamma(\infty) = \frac{\mathrm{d}r_{\mathrm{spont}}}{\alpha} \tag{3.101}$$

Integration of Equation 3.100 over a homogeneous semiconductor of finite thickness d yields

$$\mathrm{d}j_\gamma(\hbar\omega) = a(\hbar\omega)\frac{\Omega}{4\pi^3\hbar^3 c_0^2}\frac{(\hbar\omega)^2\mathrm{d}\hbar\omega}{\exp\{[\hbar\omega - (\varepsilon_{\mathrm{FC}} - \varepsilon_{\mathrm{FV}})]/kT\} - 1} \tag{3.102}$$

where the integration gives

$$a(\hbar\omega) = \left[1 - r(\hbar\omega)\right]\left[1 - \exp(-\alpha d)\right] \tag{3.103}$$

for the absorptance of a homogeneous semiconductor of thickness d equal to its emittance. Equation 3.102 is the photon current emitted into the solid angle Ω outside the semiconductor after accounting for reflection at the surface. To determine the photon current density emitted by a real solar cell, the difference of the Fermi energies in Equation 3.99 must be known as a function of position to be able to perform the integration of Equation 3.100.

Equation 3.102 is a generalization of Kirchhoff's and Planck's laws and is valid for materials that are neither black nor have a single Fermi distribution over all states. It reduces to Planck's original emission law for a thermal emitter, in which all electrons belong to a single Fermi distribution, for which $\varepsilon_{\mathrm{FC}} - \varepsilon_{\mathrm{FV}} = 0$. The difference in the Fermi energies $\mu_{\mathrm{eh}} = \varepsilon_{\mathrm{FC}} - \varepsilon_{\mathrm{FV}}$ is the free energy per electron–hole pair, also called the *chemical potential* of electron–hole pairs. It is free of entropy and we may therefore hope to transfer it into electrical energy

1) The spontaneous emission rate in (3.99) was derived for direct transitions. It is valid for indirect transitions as well, the derivation is a little more complicated because of the participation of phonons [4].

without loss. Electron–hole pairs that recombine generate photons, which are emitted. These photons carry the free energy of the electron–hole pairs, and $\mu_\gamma = \mu_{eh} = \varepsilon_{FC} - \varepsilon_{FV}$ is recognized as the chemical potential of the photons.

It is often argued that the chemical potential of the photons is $\mu_\gamma = 0$ by nature [5], because their number n_γ is not conserved. According to this argument a minimum of the free energy F could only exist if in $dF = \cdots + \mu_\gamma\, dn_\gamma + \cdots = 0$ the chemical potential $\mu_\gamma = 0$, since nonconservation of n_γ would allow $dn_\gamma \neq 0$. This argument can hardly be correct. Photons carry other quantities like energy, momentum, and angular momentum, which are strictly conserved. Furthermore, photons do not react with each other and a photon cannot split into two by itself. Changes in the photon number are only possible in reactions with matter. The involvement of other particles, electrons or phonons, ensures the conservation of energy, momentum, and angular momentum when the photon number changes. The annihilation of a photon by the generation of an electron–hole pair with conservation of energy and momentum, discussed in this chapter, is an example. We know other cases where the number of particles is not conserved without resulting in a zero chemical potential. It is rather characteristic for chemical reactions in general that the number of the particles involved is not conserved. As an example, we look at the reaction of hydrogen with oxygen to form water

$$H_2 + \tfrac{1}{2}O_2 \longleftrightarrow H_2O$$

Equilibrium exists if

$$dF = \cdots + \mu_{H_2} dN_{H_2} + \mu_{O_2} dN_{O_2} + \mu_{H_2O} dN_{H_2O} + \cdots = 0$$

With the changes in the particle numbers in the reaction $dN_{H_2} = 2dN_{O_2} = -dN_{H_2O}$ we find the equilibrium condition

$$\mu_{H_2} + \tfrac{1}{2}\mu_{O_2} = \mu_{H_2O}$$

We conclude that nonconservation of the particle numbers does not lead to zero chemical potentials if a well-defined relation among the particle numbers exists as in every chemical reaction.

The absorption and emission processes of photons are chemical reactions as well, in which electron–hole pairs are generated or annihilated. The chemical reaction is

$$e + h \rightleftharpoons \gamma$$

By the same argument as for the hydrogen–oxygen reaction, equilibrium between the electron–hole pairs and the emitted photons exists if

$$\eta_e + \eta_h = \mu_e + \mu_h = \varepsilon_{FC} - \varepsilon_{FV} = \mu_{eh} = \mu_\gamma$$

This equilibrium requires frequent interactions between electron–hole pairs and photons, i.e. frequent absorption and emission as will be present if the semiconductor is in a perfectly reflecting cavity and the temperature and chemical potential of its electron–hole pairs are kept constant (we will learn later how this is achieved).

This equilibrium exists also in a homogeneously excited semiconductor that is thicker than the penetration depth of the photons, where photons are repeatedly emitted and reabsorbed before they reach the surface.

It must be emphasized that the validity of the generalized Planck equation (3.99) and (3.102) is not restricted to a situation of equilibrium between electron–hole pairs and photons. This equilibrium was not required in its derivation. The only condition for its validity is that the Fermi distributions for the energy ranges between which optical transitions occur (e.g. conduction band and valence band) must be well defined, which is the case if the recombination lifetimes of electrons and holes are much longer than the scattering times.

The above treatment is valid quite generally, even if $\varepsilon_{FC} - \varepsilon_{FV} > \hbar\omega$, where the denominator in Equations 3.99 and 3.102 is negative. Under the same condition, the absorption coefficient in Equation 3.97 is negative too, and the spontaneous emission rate in Equations 3.98 and 3.99 remains positive. When the absorption coefficient is negative, stimulated emission overcompensates the rate of upward transitions and the semiconductor amplifies the incident light as in a laser. Amplification results from a negative absorption coefficient, which leads to a negative absorptance a, increasing exponentially with the thickness d in Equation 3.103. $\varepsilon_{FC} - \varepsilon_{FV} > \hbar\omega$ is also known as the *condition for lasing*. An equilibrium between the electron–hole pairs and the photons cannot exist if $\varepsilon_{FC} - \varepsilon_{FV} > \hbar\omega$ because then the emission rate of the semiconductor is always larger than the rate of photon absorption, since the semiconductor is a photon current amplifier under this condition.

Equation 3.99 describes the rate of radiative transitions. In a real material, there are also nonradiative transitions occurring in parallel and in addition. This does not invalidate Equation 3.99 as long as the assumptions are justified that two separate Fermi distributions describe the occupation of states in the two energy ranges between which the transitions occur.

The generalized Planck law in Equation 3.99 is important for solar cells, because it allows one to determine the smallest, theoretically possible recombination rate, the longest possible lifetime, and the largest possible difference of the Fermi energies. But the generalized Planck law has practical importance as well.

Since Equation 3.102 is a quantitative law with no adjustable parameters, it allows the experimentalist to determine the difference in the Fermi energies from an absolute measurement of the emitted photon current and a knowledge of the absorptance $a(\hbar\omega)$. The knowledge of the difference in the Fermi energies allows a better characterization of a solar cell material than does the lifetime of electrons and holes.

Alternatively, the absorptance and from it the absorption coefficient can be obtained from Equation 3.102 by measuring the emitted photon current if the difference in the Fermi energies is known. Measuring the emitted photon current density in absolute units is not easy experimentally. If the difference in the Fermi energies is not known but the emitted photon current is known except for a constant instrumental factor, the measured photon current can be adjusted to Equation 3.102, if the absorptance is known at one single photon energy allowing one to obtain the absorptance $a(\hbar\omega)$ at all other photon energies in the measured luminescence spectrum.

3.8 Problems

3.1 What are the consequences of doping of a semiconductor?

3.2 Doping of a semiconductor with donors increases the concentration of electrons. Why does this reduce the hole concentration in the dark?

3.3 What is the condition for the validity of the quasi-Fermi energy concept?

3.4 Under which conditions is the occupation of impurity states given by one of the quasi-Fermi energies even if its energy level ε_{imp} is in between the two quasi-Fermi energies? Consider different capture cross sections for electrons and holes.

3.5 What is the difference between the temperature dependences of the electrical conductivity of a metal and of an intrinsic semiconductor?

Explain qualitatively how doping changes this temperature dependence for a semiconductor.

3.6 Why is the conduction band of a semiconductor not totally unoccupied at room temperature? Would that not minimize its energy?

3.7 The density of states for electrons in the conduction band was derived in Equation 3.7. How does the density of states depend on energy for a one-dimensional and a two-dimensional electron gas?

3.8 Which assumption leads to the dependence $\alpha(\hbar\omega) \sim (\hbar\omega - \varepsilon_G)^{1/2}$ of the absorption coefficient α on the photon energy $\hbar\omega$?

3.9 Determine the minimum difference $\Delta\varepsilon$ between the electron quasi-Fermi energy and the conduction band edge of a semiconductor in units of kT for which the concentration of electrons in the conduction band n_e calculated with the Boltzmann distribution approximates the result using Fermi–Dirac statistics within 1%

(a) for $T = 300$ K.

(b) for $T = 400$ K.

3.10 Determine quantitatively the free energy for electrons and holes per volume

$$F^* = F_e^* + F_h^* = \frac{[E_e + E_h - T(S_e + S_h)]}{V}$$

in intrinsic silicon ($N_C = 3 \times 10^{19}$ cm^{-3}, $N_V = 1 \times 10^{19}$ cm^{-3} and $\varepsilon_G =$ 1.12 eV) for $T = 300$ K. Set the zero point of F^* to the state where the conduction band is empty and the valence band is fully occupied. For which concentration of n_e does F^* reach its minimum?

(a) Use the Sackur Tetrode equation (3.35) for the determination of the entropy S.

(b) Calculate the entropy according to $S = k \ln \Omega$, with Ω being the number of realizations. The number of realizations is the number of possibilities to distribute n particles in N states. Assume the density of states in the conduction band to depend on energy according to Equation 3.7

$$D_e(\varepsilon_e) = 4\pi \left(\frac{2m_e^*}{h^2}\right)^{3/2} (\varepsilon_e - \varepsilon_C)^{1/2}$$

and for the valence band according to

$$D_h(\varepsilon_h) = 4\pi \left(\frac{2m_h^*}{h^2}\right)^{3/2} (\varepsilon_h + \varepsilon_V)^{1/2}$$

3.11 Calculate the binding energy of the ground state and the first two excited states (hydrogen atom approximation) of an electron of a donor with five valence electrons in a silicon lattice. Use a dielectric function ε (Si) = 11.9 and an effective mass of

(a) $m_e^* = 1.08\ m_e$.

(b) $m_e^* = 0.02\ m_e$.

(c) How large is the Bohr radius of the ground states for case (a) and (b) and over how many atoms is the electron "smeared out"? Use $\rho_{Si} = 2.3$ g cm^{-3} and a molar mass of $M_{Si} = 28.1$ g/mol.

3.12 Calculate the Fermi energy in n-doped Si with $n_D = 10^{16}$cm^{-3} for T = 300 K,

(a) assume full ionization of the donors.

(b) numerically, use $\varepsilon_C - \varepsilon_D = 0.2$ eV.

(c) At which temperature is $n_h \geq 0.5\ n_e$? What fraction of n_e originates from the donors at that temperature?

3.13 A sample of n-doped Si is illuminated resulting in $n_e = 10^{17}$ cm^{-3} and $n_h = 10^{10}$ cm^{-3}, which are assumed to be constant throughout the sample due to large diffusion lengths. Consider acceptor-type impurity states with $\tau_{h,\min} = \tau_{e,\min} = 10^{-6}$ s. Calculate the recombination rates as a function of the impurity level ($\varepsilon_V \leq \varepsilon_{imp} \leq \varepsilon_C$) for temperatures $T = 300$ K and $T = 400$ K. Plot a graph of the recombination rates as a function of the impurity level.

4
Conversion of Thermal Radiation into Chemical Energy

In Chapter 2 we have seen what a Carnot engine can do, which operates with the heat gained from a black or a monochromatic absorber of solar radiation. The absorber was characterized by a high temperature T_A and a chemical potential of its electron–hole excitations of $\mu_{eh} = 0$. A solar cell remains at about $T_0 = 300$ K and if it had $\mu_{eh} = 0$ as well, it would not be able to convert heat into an entropy-free form of energy. But as was shown in Chapter 3, electron–hole pairs in an illuminated semiconductor have a chemical potential of $\mu_{eh} \neq 0$, which means that conversion of solar heat into chemical energy has already taken place in the semiconductor.

In order to examine this conversion in more detail, we will follow the generation and thermalization of electrons and holes step by step. These steps are shown in Figure 4.1. In the first step, photons absorbed from fully concentrated solar radiation establish an energy distribution of the electron–hole pairs, which is identical to the energy distribution of the absorbed photons reflecting the high temperature of the Sun, T_S. In this state, radiative recombination of electron–hole pairs results in the emission of photons with the same spectrum and filling (for maximal concentration) the same solid angle as the absorbed photons. This will not be changed, when, in a next step, scattering of electrons and holes between each other is allowed (the interaction with the lattice vibrations is still switched off), because scattering preserves the number of electrons and holes and the average energy per electron–hole pair. The temperature of the electron–hole pairs defined by their energy distribution is still $T_{eh} = T_S$. In this situation, the emitted photons and the electron–hole pairs in the semiconductor are in thermal and chemical equilibrium with the solar radiation and have a chemical potential of $\mu_\gamma = \mu_{eh} = 0$.

When concentration is not at maximum and emitted photons cover a larger solid angle than the absorbed photons, the emitted photon current density per solid angle is smaller than the absorbed, but still has the same spectrum. In this situation, $T_{eh} = T_S$ because of the spectrum but $\mu_\gamma = \mu_{eh} < 0$.

Returning to maximum concentration, the electrons and holes are in thermal and chemical equilibrium with the Sun with no difference of the Fermi energies. In

Physics of Solar Cells: From Basic Principles to Advanced Concepts. Peter Würfel

ISBN: 978-3-527-40857-3

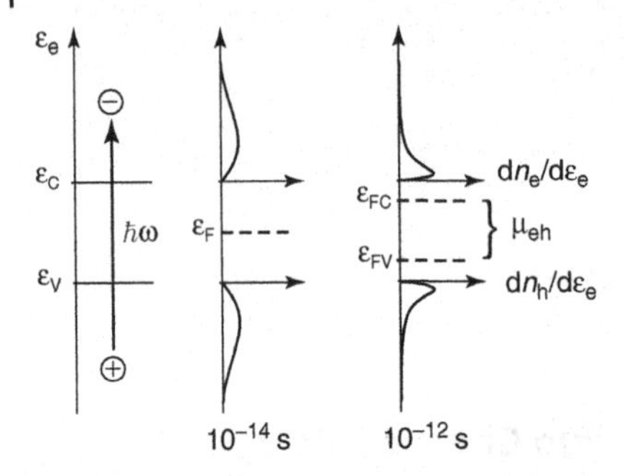

Figure 4.1 Electrons and holes generated by photons with energy $\hbar\omega$ lose energy by thermalization, which produces chemical energy per electron–hole pair μ_{eh}.

a rough estimate, the average energy of the electron–hole pairs is then $\varepsilon_G + 3kT_S$. From Equation 3.29 it follows that at $T_{eh} = T_S$

$$\mu_{eh}(T_S) = \varepsilon_{FC} - \varepsilon_{FV} = \varepsilon_G - kT_S \ln\left(\frac{N_C N_V}{n_e n_h}\right) = 0$$

from which $kT_S \ln[N_C N_V/(n_e n_h)] = \varepsilon_G$.

In the next step, the interaction with the lattice vibrations is switched on. This leads to a cooling of the electrons and holes, until after about 10^{-12} s the lattice temperature T_0 is reached, while the concentrations of electrons and holes remain constant. For $T_{eh} = T_0$

$$\mu_{eh}(T_0) = \varepsilon_{FC} - \varepsilon_{FV} = \varepsilon_G - kT_0 \ln\left(\frac{N_C N_V}{n_e n_h}\right)$$

Because of the constant concentrations of electrons and holes and neglecting the temperature dependence of the effective densities of states N_C and N_V, this finally gives us an estimate of the chemical energy per electron–hole pair produced by the cooling process

$$\mu_{eh}(T_0) = \varepsilon_{FC} - \varepsilon_{FV} = \varepsilon_G \left(1 - \frac{T_0}{T_S}\right)$$

This relation looks like a Carnot efficiency, but it is different. For the efficiency, we have to compare $\varepsilon_{FC} - \varepsilon_{FV}$ not with ε_G but with the energy $\varepsilon_G + 3kT_S$ per electron–hole pair before the cooling. We can then see a large energy loss resulting from thermalization, which can be attributed to the production of entropy by the lattice vibrations. The cooling of electrons and holes at constant concentrations is therefore far from an ideal process for the production of chemical energy, if we start from a broad energy distribution. Although in this consideration no electron–hole pairs have been extracted from the semiconductor absorber, an equilibrium between the Sun and the semiconductor does not exist if the semiconductor is at a different temperature. A nonzero chemical potential of its electron–hole pairs reduces the entropy production compared with a zero chemical potential absorber, as it enables

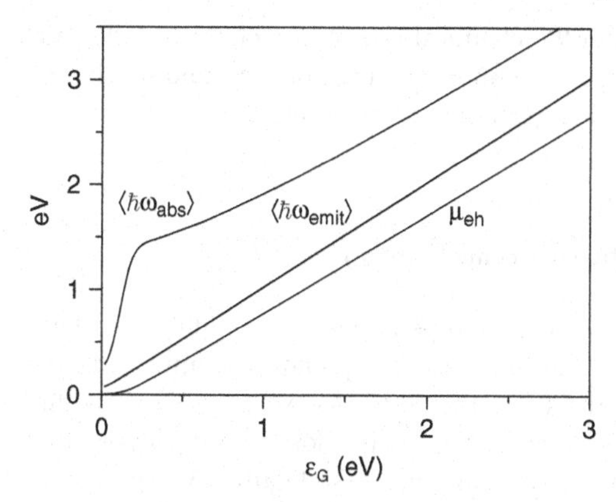

Figure 4.2 Average energy of absorbed photons $\langle \hbar\omega_{abs} \rangle$ from a nonconcentrated *AM0* spectrum and from the 300 K background radiation, the average energy of emitted photons $\langle \hbar\omega_{emit} \rangle$ and the chemical potential of electron–hole pairs μ_{eh} of a semiconductor as a function of its band gap ε_G for radiative recombination under open-circuit conditions, resulting from the generalized Planck radiation law. The difference between $\langle \hbar\omega_{abs} \rangle$ and $\langle \hbar\omega_{emit} \rangle$ is lost by thermalization of the electron–hole pairs.

the absorber to emit as many photons as it absorbs, but it does not fully compensate the temperature difference, because the emitted spectrum is different from the absorbed spectrum, and less energy is emitted than is absorbed.

Figure 4.2 shows the exact result of the thermalization of electron–hole pairs generated by nonconcentrated sunlight and by 300 K background radiation under open-circuit conditions, if there is only radiative recombination. Applying the generalized Planck law from Equation 3.102, the comparison of the average energy $\langle \hbar\omega_{abs} \rangle$ of the absorbed photons, which is invested for the generation of each electron–hole pair with the average energy $\langle \hbar\omega_{emit} \rangle$ of the emitted photons, shows the energy loss by thermalization, which, however, results in the chemical energy per electron–hole pair μ_{eh}. This chemical energy can, e.g., initiate further chemical reactions as in photosynthesis, where a long-term storage of energy is achieved. In a solar cell, the chemical energy is an intermediate product from which electrical energy can be obtained by a further step.

The chemical energy per electron–hole pair is the sum of the electrochemical potentials of the electrons and the holes, and therefore

$$\eta_e + \eta_h = \mu_e + \mu_h = \varepsilon_{FC} - \varepsilon_{FV}$$

According to Equation 3.29, this gives us

$$\mu_e + \mu_h = \varepsilon_{FC} - \varepsilon_{FV} = kT \ln\left(\frac{n_e n_h}{n_i^2}\right) \tag{4.1}$$

If we again separate the particle densities into dark-state and additional densities, so that $n_e = n_{e,0} + \Delta n_e$ and consider the case of a p-type semiconductor with $n_h \approx n_h^0$ (weak excitation), we then find

$$\mu_e + \mu_h = kT \ln\left(1 + \frac{\Delta n_e}{n_e^0}\right) \tag{4.2}$$

This means that for weak excitation the chemical energy per electron–hole pair is the result of only the change in the chemical potential of the minority charge carriers (here the electrons) or the change in their Fermi energy.

4.1
Maximum Efficiency for the Production of Chemical Energy

Is it imaginable that a material at room temperature is in equilibrium with the Sun, in which it emits toward the Sun as many photons, as much energy and as much entropy as it absorbs from the Sun? As we have seen in the previous paragraph, the main problem is the energy loss accompanying the entropy production during the thermalization process. Figure 4.3 shows that an energy loss cannot occur if the states available for electrons and holes are confined to a narrow energy interval. Such a material could only absorb almost monochromatic radiation. Alternatively, if in a broadband semiconductor the electrons and holes are generated by almost monochromatic radiation, and only into those states in which they are after the thermalization, an energy loss can be avoided. For maximum efficiency, we therefore make the following idealized assumptions:

1. Only radiative recombination takes place.
2. There is no extraction of electrons and holes.
3. There are no thermalization losses. Absorption and emission are monochromatic, with an energy of $\hbar\omega = \varepsilon_G$ and an absorptance of $a(\hbar\omega = \varepsilon_G) = 1$ over an interval $d\hbar\omega$. We can consider this condition fulfilled by a filter enclosing the semiconductor, which is transparent only for $\hbar\omega = \varepsilon_G$, reflecting all other photons.
4. The rate of generation is at maximum. This condition is achieved by maximum concentration, where the solid angles for absorption Ω_{abs} and emission Ω_{emit} are equal.

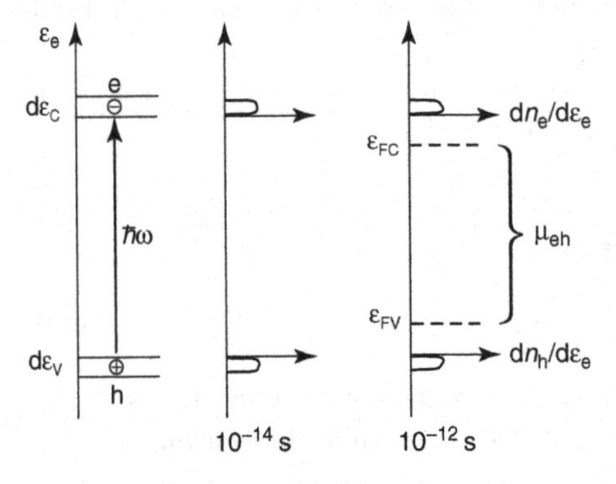

Figure 4.3 Electrons and holes generated in narrow energy ranges by monochromatic radiation have the same energy distribution after thermalization as before.

From the first two conditions, it follows that in the steady state the emitted photon current is equal to the absorbed photon current. Referring to Equation 2.35 and Equation 3.102 this means that

$$\begin{aligned} dj_{\gamma,\,emit} &= \frac{\Omega_{emit}}{4\pi^3\hbar^3c^2}\frac{\varepsilon_G^2\,d\hbar\omega}{\exp\left(\dfrac{\varepsilon_G-\mu_{eh}}{kT_0}\right)-1} \\ &= \frac{\Omega_{abs}}{4\pi^3\hbar^3c^2}\frac{\varepsilon_G^2\,d\hbar\omega}{\exp\left(\dfrac{\varepsilon_G}{kT_S}\right)-1} = dj_{\gamma,abs} \end{aligned} \tag{4.3}$$

With condition 4, $\Omega_{emit} = \Omega_{abs}$, under which the semiconductor would reach the temperature of the Sun T_S unless we hold its temperature constant at $T = T_0$, we find the chemical energy per electron–hole pair

$$\mu_{eh} = \mu_e + \mu_h = \varepsilon_G\left(1 - \frac{T_0}{T_S}\right) \tag{4.4}$$

The efficiency for the conversion of solar heat into chemical energy is

$$\eta = \frac{\mu_e + \mu_h}{\varepsilon_G} = 1 - \frac{T_0}{T_S} \tag{4.5}$$

This is the Carnot efficiency, a limiting value that is obtained for the conversion of heat into an entropy-free form of energy when the conversion process is reversible, i.e., takes place without the production of entropy. We can now see that an ideal semiconductor, which only has radiative recombination, represents an ideal converter of heat into chemical energy in monochromatic operation.

The state defined by Equation 4.4 is a state of equilibrium between the semiconductor and the Sun, in which not only are the absorbed and emitted photon currents equal, as expressed by Equation 4.3, but also the energy currents that follow from the photon currents by multiplication with the photon energy, which for absorption and emission have the same value $\hbar\omega = \varepsilon_G$. For the absorbed and emitted radiation, the entropy per photon state is [6]

$$\sigma_\gamma = k\left[(1+f_\gamma)\ln(1+f_\gamma) - f_\gamma \ln(f_\gamma)\right] \tag{4.6}$$

where f_γ is the Bose–Einstein distribution function

$$f_\gamma = \frac{1}{\exp\left[(\hbar\omega - \mu_\gamma)/kT\right] - 1} \tag{4.7}$$

The exponent in the distribution function has the same value for the Sun and the semiconductor according to Equations 4.3 and 4.4 with $\mu_e + \mu_h = \mu_\gamma$ because of the chemical equilibrium between the electron–hole pairs and the photons by frequent absorption and emission in a material with absorptance $a(\hbar\omega = \varepsilon_G) = 1$. In addition, absorption and emission involve the same number of photon states because of $\Omega_{abs} = \Omega_{emit}$. Under these conditions, the entropy absorbed with the photons from the Sun is equal to the entropy emitted with the photons toward the Sun. This all shows that an equilibrium is possible between the Sun and a material at room temperature, if the absorption and emission of photons is limited to the same narrow energy range. This equilibrium requires the right combination of temperature T and chemical potential μ as expressed by Equation 4.4, which is why we may call it a thermochemical equilibrium. The chemical energy produced by reversible, nondissipative cooling of electrons and holes is carried away by the emitted photons. The emitted photons, known as *luminescent radiation* from LEDs, look the same as the solar photons. For monochromatic radiation, there is no way to decide whether they have a high temperature T and a small chemical potential μ_γ or a low temperature and a large chemical potential.

The process in which all the chemical energy is emitted with the photons is in fact of no interest to us. We are instead interested in how much chemical energy can be harvested with the electron–hole pairs, if we knew how to extract them from the absorber. From the continuity equation for the electrons under steady-state conditions,

$$\frac{\partial n_e}{\partial t} = G_e - R_e - \mathrm{div}\, \mathrm{d}j_e = 0$$

we see that

$$\mathrm{div}\, \mathrm{d}j_e = G_e - R_e \qquad (4.8)$$

gives just the rate with which electrons can be extracted from a volume element, since more flow out of this element than into it when $G_e > R_e$. We are considering a differentially small electron current $\mathrm{d}j_e$, since it results from absorption and emission in a differentially small photon energy interval $\mathrm{d}\hbar\omega$.

Integrating Equation 4.8 over the volume of the semiconductor gives the total extracted electron current $\mathrm{d}j_e$

$$\mathrm{d}j_e = \int G_e\, \mathrm{d}x - \int R_e\, \mathrm{d}x$$

If we again restrict ourselves to radiative recombination, the extracted electron current density is given by the difference between the absorbed and emitted photon current densities, which already includes photon recycling by reabsorption. Thus,

$$\mathrm{d}j_e = \mathrm{d}j_{\gamma,\mathrm{abs}} - \mathrm{d}j_{\gamma,\mathrm{emit}} \qquad (4.9)$$

The same relationship is true for the holes. The current of extracted electron–hole pairs is $\mathrm{d}j_{eh} = \mathrm{d}j_e = \mathrm{d}j_h$.

For a spatially constant chemical potential of the electron–hole pairs $\mu_{eh} = \mathrm{const}$, j_{eh} follows from the relations in Equation 4.3 for the absorbed and emitted photon currents.

In an approximation to the generalized Planck law, when $\mu_{eh} < \varepsilon_G - 3kT$ allows us to neglect the "−1" in the denominator of the Planck law, we find according to Equation 4.1

$$\mathrm{d}j_{\mathrm{eh}} = \mathrm{d}j_{\gamma,\mathrm{abs}} - \mathrm{d}j_{\gamma}^{0} \frac{n_{\mathrm{e}} n_{\mathrm{h}}}{n_{\mathrm{i}}^{2}} = \mathrm{d}j_{\gamma,\mathrm{abs}} - \mathrm{d}j_{\gamma}^{0} \exp\left(\frac{\mu_{\mathrm{e}} + \mu_{\mathrm{h}}}{kT}\right) \tag{4.10}$$

where $\mathrm{d}j_{\gamma}^{0}$ is the photon current density absorbed and emitted in equilibrium with the 300 K background radiation.

The condition $\mu_{\mathrm{e}} + \mu_{\mathrm{h}} = \text{constant}$ is fulfilled, for example, when electrons and holes are produced uniformly throughout the volume as a result of weak absorption or when they are uniformly distributed throughout the volume as a result of large diffusion constants and lifetimes.

For a given current of absorbed photons $\mathrm{d}j_{\gamma,\mathrm{abs}}$, the current $\mathrm{d}j_{\mathrm{eh}}$ of extracted electrons and holes is the difference between the absorbed and the emitted photon currents, which are shown in Figure 4.4 as a function of the chemical energy extracted along with the electron–hole current. We see that the extracted electron–hole current is nearly equal to the absorbed photon current for small values of $\mu_{\mathrm{eh}} = \mu_{\mathrm{e}} + \mu_{\mathrm{h}}$, where almost no photons are emitted. The emitted photon current $\mathrm{d}j_{\gamma,\mathrm{emit}}$ rises exponentially with $\mu_{\mathrm{e}} + \mu_{\mathrm{h}}$, until the open-circuit situation is reached at $\mu_{\mathrm{eh,oc}}$, where absorbed and emitted photon currents are equal and no electron–hole pairs are extracted. The current of chemical energy extracted along with the electron–hole pairs is

$$\mathrm{d}j_{\mu} = \mathrm{d}j_{\mathrm{eh}}\,(\mu_{\mathrm{e}} + \mu_{\mathrm{h}})$$

Its maximum value is the largest rectangle (dark grey in Figure 4.4), which can be fitted between the absorbed and emitted photon currents. The absorbed energy current in the interval $\mathrm{d}\hbar\omega$ is $\mathrm{d}j_{\gamma\mathrm{abs}}\,\hbar\omega$ and is given by the lightly shaded rectangle with the broken-line border in Figure 4.4. The efficiency with which chemical energy is obtained from the absorbed radiation is finally given by the ratio of the dark and lightly shaded rectangles.

Figure 4.5 shows the efficiency η with which chemical energy is extracted as a function of the chemical potential of the electron–hole pairs. It starts with zero

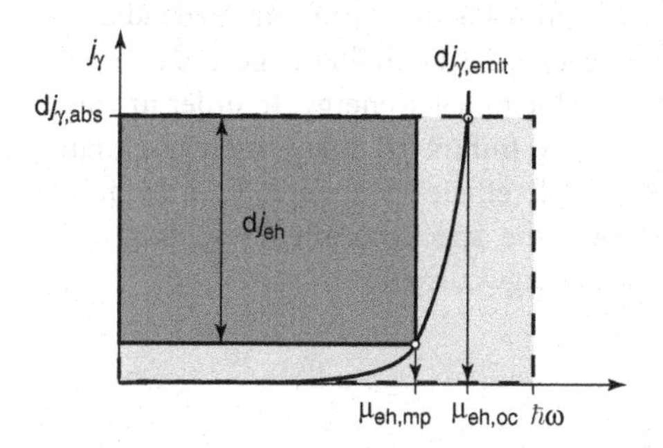

Figure 4.4 Current $\mathrm{d}j_{\gamma,\mathrm{emit}}$ of emitted photons as a function of the chemical energy $\mu_{\mathrm{eh}} = \mu_{\mathrm{e}} + \mu_{\mathrm{h}}$ of the electron–hole pairs. The dark rectangle is the current of chemical energy extracted along with the current $\mathrm{d}j_{\mathrm{eh}} = \mathrm{d}j_{\gamma,\mathrm{abs}} - \mathrm{d}j_{\gamma,\mathrm{emit}}$ of extracted electron–hole pairs. It is produced from the current of absorbed energy represented by the grey rectangle.

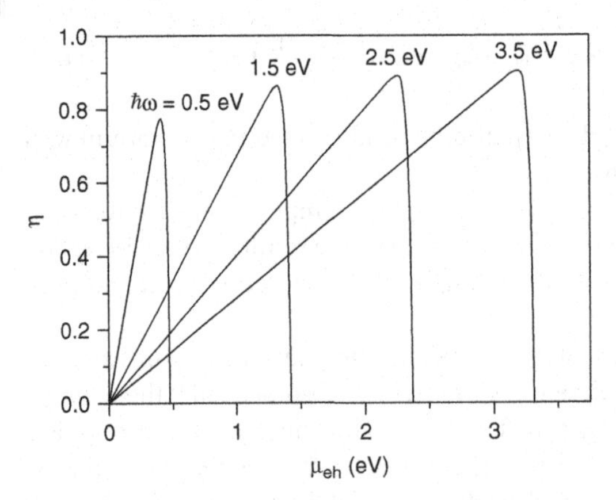

Figure 4.5 Efficiency η with which chemical energy is extracted as a function of the chemical potential μ_{eh} of the electron–hole pairs from monochromatic absorbers for different photon energies.

at $\mu_{eh} = 0$, when electrons and holes are so effectively extracted that none can accumulate, then it rises linearly and reaches a maximum at the point of maximum power (mp), after which it drops sharply down to zero at the open-circuit situation, where all electron–hole pairs recombine to emit photons.

In order to find the maximum efficiency with which chemical energy can be obtained from fully concentrated solar radiation, we proceed as follows. From the generalized Planck law in Equation 3.102, the absorbed and emitted photon currents are found as a function of the monochromatic photon energy. For each photon energy, the chemical potential μ_{eh} in the emitted photon current is varied until the value $\mu_{eh,mp}$ is found for which the current of extracted chemical energy is maximal. Dividing this chemical energy current by the absorbed monochromatic energy current gives the efficiency $\eta_{mono}(\varepsilon_G)$ as a function of the band gap.

The result of this calculation is shown in Figure 4.6 for nonconcentrated radiation and for fully concentrated radiation. The efficiencies are rather large and rise with the band gap ε_G, demonstrating the high value of solar energy. In order to make proper use of the total solar spectrum, very (infinitely) many monochromatic absorbers have to be employed. The overall efficiency then follows from an integration of the efficiency $\eta_{mono}(\varepsilon_G)$ over the spectrum after weighting each interval $d\hbar\omega$ with its share of the absorbed energy current.

$$\eta = \frac{\int_0^\infty \eta_{mono}(\varepsilon_G)\,\hbar\omega\,dj_{\gamma,abs}}{\int_0^\infty \hbar\omega\,dj_{\gamma,abs}}$$

The total efficiency is 86% for fully concentrated radiation from a 5800 K Sun and 67% for nonconcentrated radiation from a solid angle of 6.8×10^{-5} [7].

Although we were able to calculate the efficiency of the chemical energy that can be obtained from an illuminated semiconductor by the extraction of some of the

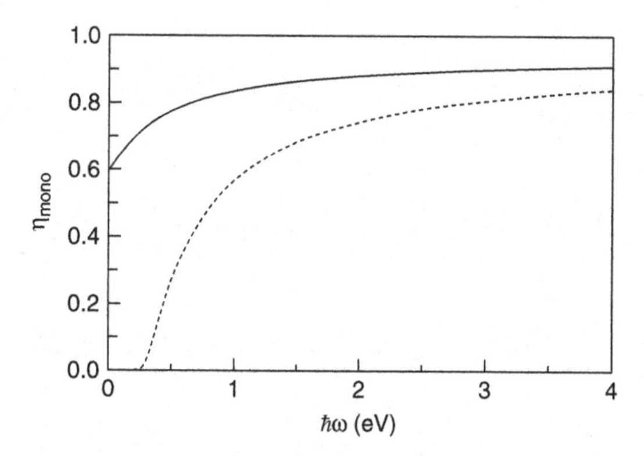

Figure 4.6 Monochromatic efficiency $\eta_{mono}(\varepsilon_G)$ for obtaining chemical energy $\mu_e + \mu_h$ as a function of the photon energy $\hbar\omega$ of fully concentrated (full line) and nonconcentrated (broken line) monochromatic solar radiation.

generated electron–hole pairs, we still have no idea how a device should look that allows this extraction of electron–hole pairs and their energy. This is the subject of the next chapter.

4.2 Problems

4.1 What is the reason for the linear increase of the efficiency η of the extraction of chemical energy in Figure 4.5 for small values of μ_{eh}?

4.2 Calculate the quasi-Fermi energy of the electrons in a p-type Si sample ($n_h \approx n_h^0 = 5 \times 10^{16}\ \text{cm}^{-3}$) at $T_0 = 300$ K under open-circuit conditions, regarding

(a) only radiative recombination;

(b) only Auger recombination.

Assume excitation by black-body radiation with $T_{bb} = 5800$ K, from a solid angle of $\Omega = 6.8 \times 10^{-5}$ and an absorptance $a(\hbar\omega < \varepsilon_G) = 0$ and $a(\hbar\omega \geq \varepsilon_G) = 1$ with $\varepsilon_G = 1.12$ eV. The charge carriers shall be distributed homogeneously over a thickness of $d = 300\ \mu\text{m}$ because of large diffusion lengths.

4.3 Calculate the emitted photon current density per wavelength into an effective solid angle of π for $\lambda_G = 500$ nm. At this wavelength, the absorber is characterized by an absorptance of $a(\lambda_G) = 1$. The emission occurs under conditions where the chemical potential of the electron–hole pairs is

(a) $\mu_{eh} = 0.5$ eV;

(b) $\mu_{eh} = 1.8$ eV.

5
Conversion of Chemical Energy into Electrical Energy

When electrons and holes are removed pairwise from the same location and along the same path, there is no charge current connected with the extraction, because an electron–hole pair is electrically neutral. Electrical energy currents are bound to charge currents. It is therefore necessary to separate the electrons and the holes and extract them along different paths as shown in Figure 5.1. Extraction, in fact, is the wrong word. Electrons and holes must rather be driven out of an absorber by some internal force, since they must still be able to perform work in an external circuit.

In this chapter, we find out which are the forces that drive electrons and holes and how they may be provided by a solar cell structure.

5.1
Transport of Electrons and Holes

Electrons have many "handles" where forces can be applied. The gradient of the gravitational potential acts on their mass, and the gradient of the electrical potential on their charge. The temperature gradient acts on their entropy, and the gradient of the chemical potential acts on their quantity. The gravitational force is negligibly small. Temperature gradients give rise to thermoelectric effects and we disregard them in connection with solar cells. Even though we already know that the charge and number of quantity are coupled during the exchange of electrons or holes, and consequently the forces acting on them are coupled, we first treat the effects of these forces separately. For this, we assume that only one of the two forces (gradient of the electrical potential and gradient of the chemical potential) is different from zero. At the same time, we must remember that the forces acting on the charge and those acting on the quantity, act on the same particles and must therefore be added to obtain a resultant force, which is then responsible for driving the particles.

Physics of Solar Cells: From Basic Principles to Advanced Concepts. Peter Würfel

ISBN: 978-3-527-40857-3

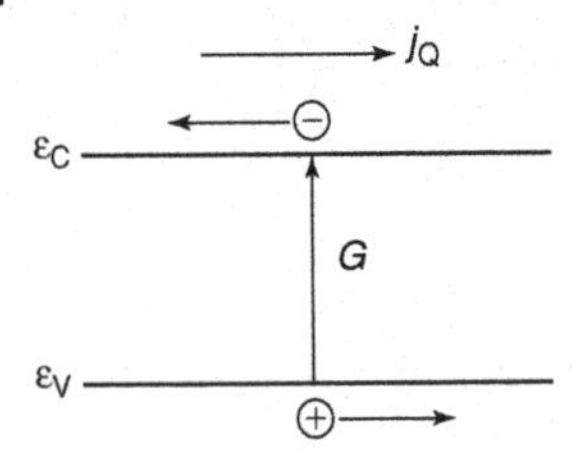

Figure 5.1 Electrons in the conduction band and holes in the valence band have to move in different directions to produce an electrical charge current j_Q.

5.1.1 Field Current

The electric field $E = -\text{grad}\varphi$ is the driving force acting on the charge. It is the only driving force for the charge current of electrons and holes if their concentration is spatially uniform. For electrons, then, their chemical potential

$$\mu_e = \mu_{e,0} + kT \ln\left(\frac{n_e}{N_C}\right)$$

is independent of position and

$$\text{grad}\,\mu_e = 0$$

the driving force for the diffusion current, still to be discussed, is zero. Figure 5.2 illustrates the dependence on position of the electron energies. The difference between the Fermi energy ε_{FC} and the conduction band edge ε_C does not depend on position, because the concentration of electrons is assumed to be uniform throughout. The gradient of ε_C is equal to the gradient of the electrical energy $-e\varphi$ because of the uniform electron affinity $\chi_e = -\mu_{e,0}$ in a uniform material.

The density of the charge field current for particles of type i with a charge per particle of $z_i e$ is

$$j_{Q,f,i} = z_i e n_i \langle v_i \rangle \tag{5.1}$$

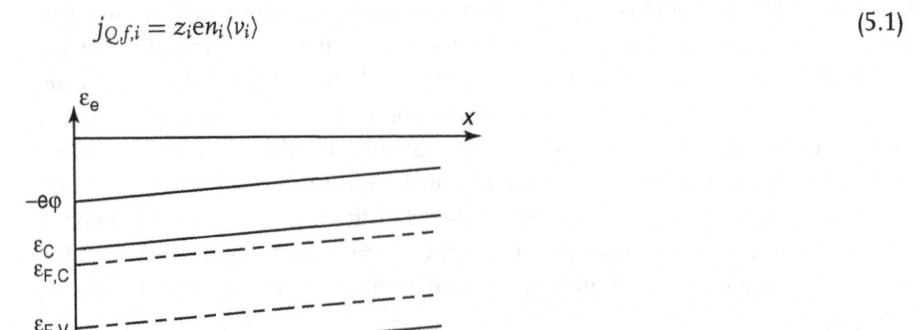

Figure 5.2 Electron energies in an electric field for uniform electron and hole concentrations.

The particles – with a concentration n_i – move at the mean velocity $\langle v_i \rangle$ and carry along the charge z_i e. Since the mean value of the velocity in the absence of an electric field, the mean thermal velocity $\langle v_{th} \rangle$, is zero, $\langle v_i \rangle$ is the mean value of the velocity in the presence of an electric field, or the drift velocity, which is small compared with $\langle v_{th} \rangle$.

Owing to their motion, which is mainly thermal, the charge carriers collide with obstacles, caused by any disturbance of the perfect periodicity of a crystal, such as phonons or impurities. The mean distance between collisions is called the *mean free path.*

The mean time $\tau_{c,i}$ between two collisions is called the *collision time.* In the electric field, the motion between collisions is characterized by the acceleration $a_i = z_i eE/m_i^*$. Because of the exponential distribution of the time intervals between collisions – a few particles undergo collisions only after a time much longer than $\tau_{c,i}$ – the mean velocity is given by

$$\langle v_i \rangle = \int_0^\infty a_i \exp(-t/\tau_{c,i})\, dt = a_i\, \tau_{c,i}$$

thus

$$\langle v_i \rangle = z_i \frac{e}{m_i^*} \tau_{c,i}\, E \tag{5.2}$$

$b_i = e\tau_{c,i}/m_i^*$ is called the *mobility of the particles of type i.* The density of the field current is then

$$j_{Q,f,i} = z_i^2 e n_i b_i E = \sigma_i E \tag{5.3}$$

Here $\sigma_i = z_i^2 e n_i b_i$ is the conductivity of the particles of type i. Making use of $E = -\text{grad}\varphi$ we can also express the charge current in terms of the gradient of the electrical energy per particle $z_i e\varphi$

$$j_{Q,f,i} = -\frac{\sigma_i}{z_i e} \text{grad}\,(z_i e\varphi) \tag{5.4}$$

For electrons with $z_i = -1$, this is

$$j_{Q,f,e} = \frac{\sigma_e}{e} \text{grad}\,(-e\varphi) \tag{5.5}$$

and for holes with $z_i = +1$,

$$j_{Q,f,h} = -\frac{\sigma_h}{e} \text{grad}\,(e\varphi) \tag{5.6}$$

5.1.2
Diffusion Current

When the electrical potential is the same at every location, that is, $\text{grad}\varphi = 0$, a pure diffusion current flows if the concentration is nonuniform. As Figure 5.3 shows, in a semiconductor made of uniform material the conduction band edge ε_C then has the same value everywhere. The difference between the Fermi energy ε_{FC} and the conduction band edge is concentration-dependent and therefore the nonuniform component of the chemical potential of the electrons

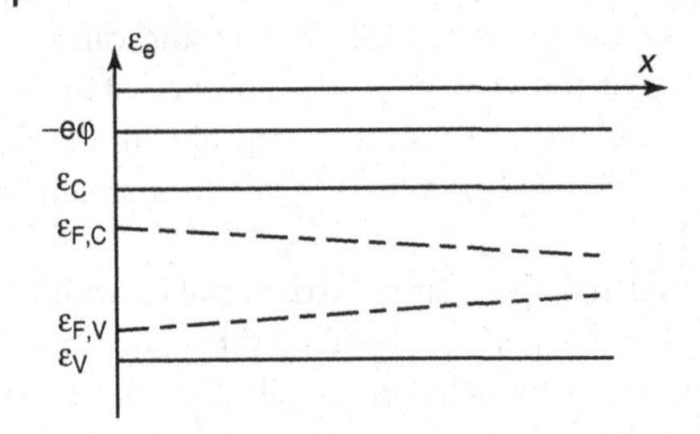

Figure 5.3 Electron energies for a position-dependent electron concentration in the absence of an electric field.

$$\mu_e - \mu_{e,0} = kT \ln \frac{n_e}{N_C}$$

where N_C is the effective density of states for electrons.

In the usual representation by Fick's Law, the charge current due to a nonuniform concentration of particles of type i is given by

$$j_{Q,d,i} = z_i e(-D_i \operatorname{grad} n_i) \tag{5.7}$$

where D_i is the diffusion coefficient. In this form, the current of the particles of type i is not proportional to their concentration n_i and thus does not allow us to recognize the driving force acting on this concentration of particles. We can make the current proportional to the concentration by transforming this expression

$$j_{Q,d,i} = z_i e n_i D_i \frac{\operatorname{grad} n_i}{n_i}$$

Substituting $(\operatorname{grad} n_i)/n_i$ by $\operatorname{grad} \ln (n_i/N_i)$ and making use of the chemical potential we then obtain

$$j_{Q,d,i} = -\frac{z_i e n_i D_i}{kT} \operatorname{grad} \mu_i$$

In contrast to Fick's law, this expression is valid for an inhomogeneous chemical environment ($\operatorname{grad} \mu_{i,0} \neq 0$) as well, which would cause a diffusion current even if the concentration were homogeneous.

Using the so-called Einstein relation between diffusion coefficient D_i and mobility b_i

$$\frac{b_i}{D_i} = \frac{e}{kT}$$

we finally arrive at

$$j_{Q,d,i} = -\frac{z_i e n_i b_i}{e} \operatorname{grad} \mu_i = -\frac{\sigma_i}{z_i e} \operatorname{grad} \mu_i \tag{5.8}$$

For electrons with $z_i = -1$, the diffusion current is given by

$$j_{Q,d,e} = \frac{\sigma_e}{e} \operatorname{grad} \mu_e \tag{5.9}$$

and for holes with $z_i = +1$, the diffusion current is

$$j_{Q,d,h} = -\frac{\sigma_h}{e} \operatorname{grad} \mu_h \tag{5.10}$$

5.1.3 Total Charge Current

We have seen that there are two driving forces acting on all electrons and two other driving forces acting on all holes. To get the currents caused by the forces, they have to be multiplied by the conductivities. If both forces, the gradient of the electrical energy and the gradient of the chemical potential are present simultaneously, they have to be added to give a resultant force, which is then multiplied by the conductivity to give the resultant current. We have written the field current and the diffusion current in such a way that the combination of the forces is easily possible. For the resultant current $j_{Q,i}$ of the particles of type i:

$$j_{Q,i} = -\frac{\sigma_i}{z_i e}\left\{\operatorname{grad}\mu_i + \operatorname{grad}(z_i e\varphi)\right\} \tag{5.11}$$

or

$$j_{Q,i} = -\frac{\sigma_i}{z_i e}\operatorname{grad}(\mu_i + z_i e\varphi) = -\frac{\sigma_i}{z_i e}\operatorname{grad}\eta_i \tag{5.12}$$

The gradient of the electrochemical potential, $\operatorname{grad}\eta_i = \operatorname{grad}(\mu_i + z_i e\varphi)$, now gives us the combination of the individual forces acting separately on the quantity of the particles and on their charge. The gradient of the electrochemical potential is the total force, which, in a general case, drives the total current, when either the concentration of the particles or their chemical environment or the electrical potential, or all three, are inhomogeneous.

This procedure may not be very common. In most books, even in the general case, the argument is based on field and diffusion currents, as though there were electrons that "feel" only the electric field and others that only contribute to the diffusion current.

Since the only charged particles moving in a semiconductor are electrons and holes, the total charge current is

$$j_Q = \frac{\sigma_e}{e}\operatorname{grad}\eta_e - \frac{\sigma_h}{e}\operatorname{grad}\eta_h \tag{5.13}$$

or, making use of the quasi Fermi energies, for which $\varepsilon_{FC} = \eta_e$ and $\varepsilon_{FV} = -\eta_h$,

$$j_Q = \frac{\sigma_e}{e}\operatorname{grad}\varepsilon_{FC} + \frac{\sigma_h}{e}\operatorname{grad}\varepsilon_{FV} \tag{5.14}$$

This relationship is always true, i.e. when either an electric field or a concentration gradient, or both, exist.

Figure 5.4 gives an example in which both an electric field and concentration gradients for electrons and holes exist. According to Equation 5.14, the charge current j_Q is zero in spite of nonzero field and diffusion forces, because the Fermi energies are uniform. A separate description of the effects of the field and diffusion forces concludes that a field current and a diffusion current both flow and that both compensate each other exactly. The result, a zero total charge current, is identical. Nevertheless, the idea behind this result, i.e. all electrons and holes produce the field current as a result of the motion induced by the field, and, at the same time,

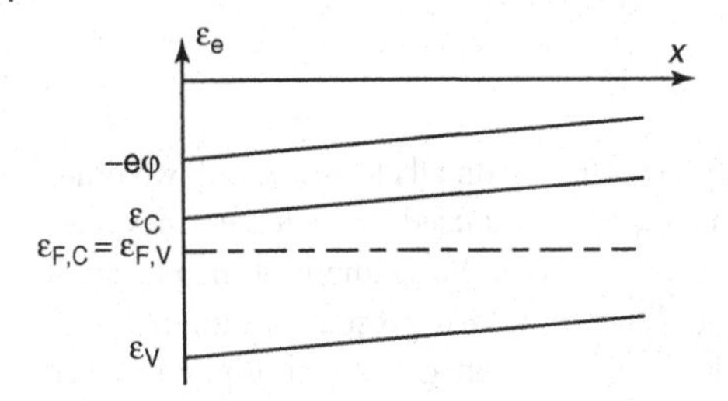

Figure 5.4 Constant Fermi energy in a semiconductor in which both an electric field and a concentration gradient exist.

all electrons and holes produce the diffusion current in the opposite direction, is incorrect.

That the field and diffusion currents do not exist separately can be seen very well if we examine the energy dissipation, that is, the generation of Joule heat associated with every current flow because of the scattering of the charge carriers. The rate of energy dissipation per volume element $d\rho_E/dt$ is proportional to the particle current density and the driving force. For the field current, it follows from Equation 5.4 that

$$\frac{d\rho_E}{dt} = \frac{j_{Q,f,i}}{z_i e}(-\text{grad}\,(z_i e\varphi)) = \frac{j_{Q,f,i}^2}{\sigma_i}$$

and for the diffusion current, it follows from Equation 5.8 that

$$\frac{d\rho_E}{dt} = \frac{j_{Q,d,i}}{z_i e}(-\text{grad}\,\mu_i) = \frac{j_{Q,d,i}^2}{\sigma_i}$$

As expected, in both cases the rate of energy dissipation $d\rho_E/dt > 0$, even if the individual currents would flow in opposite directions and compensate each other.

Figure 5.4 illustrates the case of thermal and electrochemical equilibrium of electrons and holes, as discussed in detail for the example of the pn-junction in Section 6.4. In this equilibrium state, dissipation of energy must not occur. In reality, the electrons and holes do not move in any preferred direction beyond their thermally random motion, because the forces acting on them add to a total force of $\text{grad}\,\varepsilon_F = 0$, and no energy is dissipated.

The situation is the same as for the molecules in the air. There are also two forces acting on them: the gravitational force pulling them downwards and the pressure gradient or the gradient of the chemical potential pushing them upwards. We know the result. The distribution of the molecules is such that the two forces compensate each other. There is only Brownian motion of the molecules about an equilibrium position and there are no currents.

This discussion shows that field and diffusion currents are pure fiction and do not exist separately, and only the total current representation according to Equation 5.14 is correct.

However, it may not be concluded from this discussion that a mathematical treatment of charge transport in semiconductors in terms of field and diffusion currents is incorrect. Mathematically, there is no difference, whether the driving forces are first added to give a resultant driving force, which is then multiplied by the conductivity to yield the total current, or whether the driving forces are first multiplied separately by the conductivity and then added to give the total current. The difference is in the physical picture that follows from the mathematical procedure. We obtain the physically correct description for the motion of the charge carriers only when we first add the driving forces to give a resultant driving force, which is the gradient of the Fermi energy.

5.2
Separation of Electrons and Holes

Now that we are familiar with the driving force for the motion of electrons and holes, we shall return to the original problem of defining a structure in which illumination produces a charge current, along with an electrical voltage V at its terminals, arising from a difference of the Fermi energies at the left and right terminals

$$eV = \varepsilon_{F,\text{left}} - \varepsilon_{F,\text{right}} \neq 0$$

We begin with a homogeneously exposed n-doped semiconductor. This is a structure in which, because of its symmetry, there is no preference for the transport of electrons in one direction and holes in the opposite direction, and no current or voltage is expected.

Figure 5.5 shows the Fermi energies between the bands. As a result of an assumed strong surface recombination, the concentrations of electrons and holes at the surfaces at the right and left do not differ from their values in the dark state, even though the semiconductor is illuminated. The Fermi energies for the conduction and valence bands, which are different inside the semiconductor, therefore merge into a single Fermi energy at the surface. This results in gradients for the two Fermi energies, which drive electrons and holes toward both surfaces,

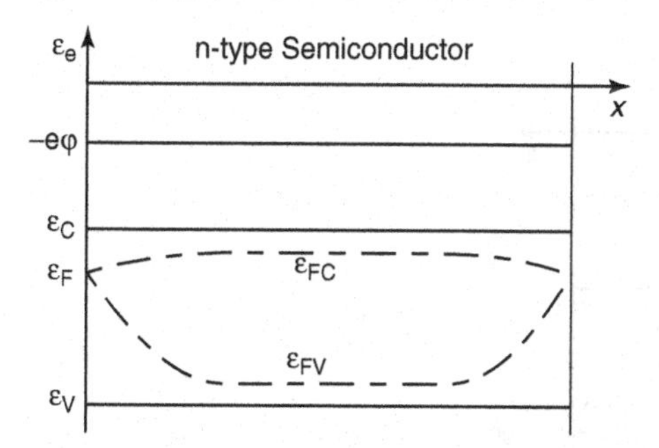

Figure 5.5 Distribution of the Fermi energies of a homogeneously illuminated n-type semiconductor with strong surface recombination on the left and on the right.

where they recombine. Since the absence of an external circuit implies that no charge current can flow, the particle currents of electrons and holes flowing to the same surface must be equal. Owing to the larger conductivity of electrons for the n-conductor under discussion, it follows from Equation 5.14 that the gradient of the Fermi energy for the conduction band ε_{FC} is smaller than the gradient of ε_{FV}. In a p-conductor this would be just the opposite. This fact leads us to conclude that replacing the n-doping in the right half of Figure 5.5 by p-doping must give the distribution of Fermi energies in Figure 5.6, for a situation where no charge current flows. We see that, in an illuminated pn configuration without charge current, i.e. under open-circuit conditions, a difference in the Fermi energies results between two surfaces with strong surface recombination, fixing the carrier concentrations at the surfaces at different values.

The electrons flow toward the left when their electrochemical potential, the Fermi energy, $\eta_e = \varepsilon_{FC}$, decreases toward the left. The holes flow toward the right when their electrochemical potential, $\eta_h = -\varepsilon_{FV}$, decreases toward the right (when ε_{FV} increases toward the right). For the electron current toward the left and the hole current toward the right the gradients of the Fermi energies ε_{FC} on the left and ε_{FV} on the right are small, since the conductivities of the electrons in the n-conductor and of the holes in the p-conductor, are large. This is the important property that causes the Fermi energies at the two surfaces to be different.

The gradients of the Fermi energies for the valence band in the n-conductor on the left and for the conduction band in the p-conductor on the right, which are unavoidable owing to the convergence of the Fermi energies at the surface, can become large. This results from the very small concentrations, and consequently small conductivities, of holes in the n-conductor and of electrons in the p-conductor. The arrangement of Figure 5.6 is, however, not an ideal converter of chemical energy into electrical energy, because the difference in the Fermi energies between the left and right surfaces is less than the separation within the semiconductor. The chemical energy per electron–hole pair ($\varepsilon_{FC} - \varepsilon_{FV}$) resulting from exposure to light cannot be fully utilized by an external circuit. This is due to too small a concentration of electrons in the n-conductor and of holes in the p-conductor in the dark state, at least near the surfaces, which require a nonnegligible gradient for the

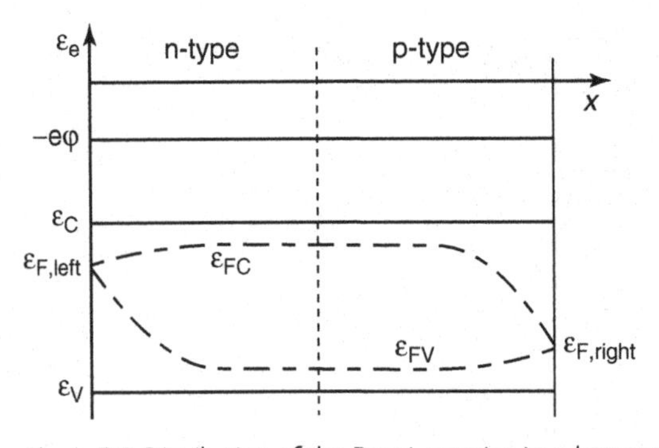

Figure 5.6 Distribution of the Fermi energies in a homogeneously illuminated pn-structure.

majority carrier Fermi energies, even under open-circuit conditions, to compensate the current of the minority carriers. A charge current flows in an external circuit, when the voltage V between the terminals is reduced below the open-circuit value. This increases the gradients of the Fermi energies for the majority carriers and decreases the gradients for the minority carriers, resulting in a preferential movement of electrons to the left and of holes to the right in Figure 5.6. We return to this point in the following chapter.

The prerequisite for the existence of the gradients of the Fermi energies required to drive the currents (without applying a voltage externally) is the separation of the Fermi energies. Since the charge current densities of the electrons and holes are, at most, equal to the current density of the photons absorbed, multiplied by the elementary charge, (in Si $\leq 42\ \mathrm{mA\ cm^{-2}}$), for the usual doping levels of $(10^{16} - 10^{17})\ \mathrm{cm^{-3}}$ in both the n-conductor and the p-conductor, only a very small Fermi energy gradient is required for the majority carriers. In spite of the current flow, a large part of the Fermi energy separation is still present as the difference of the Fermi energies at the terminals on the left and right. This produces the voltage V at the terminals of the solar cell:

$$V = \frac{1}{e} \int_{\mathrm{left}}^{\mathrm{right}} \mathrm{grad}\, \varepsilon_{FC}\, dx \tag{5.15}$$

Since $\varepsilon_{FC} = \varepsilon_{FV}$ in both terminals, one could also integrate over $\mathrm{grad}\, \varepsilon_{FV}$.

That the voltage is given by the difference between the Fermi energies on the left and right with $eV = \Delta\varepsilon_F$, results from the fact that $\mathrm{grad}\, \varepsilon_F$ is the driving force for the charge current in the external circuit as well, e. g. in a voltmeter with finite internal resistance (an infinite internal resistance does not exist). The voltmeter shows 0 V when there is no driving force for the electrons within the voltmeter, i.e. $\mathrm{grad}\, \varepsilon_F = 0$ within the voltmeter or $\Delta\varepsilon_F = 0$ at its terminals.

We see that a voltmeter, in fact, always measures differences in the electrochemical potential of the electrons if these are the only mobile particles within the instrument. A voltmeter does not measure differences $\Delta\varphi$ of the electrical potential as, e.g. between the left and right sides of the semiconductor in Figure 5.4, where it would show 0 V although $\Delta\varphi \neq 0$.

5.3 Diffusion Length of Minority Carriers

Our goal is that of the charge carriers produced by the absorption of photons, as many as possible should flow toward the terminals of the solar cell. Unfortunately, however, the existence of the driving force for the charge carriers is not sufficient. Since the electrons and holes recombine after a lifetime, they must also be able to reach the terminals in this time. For the structure shown in Figure 5.6 the transport mechanism is diffusion. There is no gradient of the electrical potential and no field present in Figure 5.6.

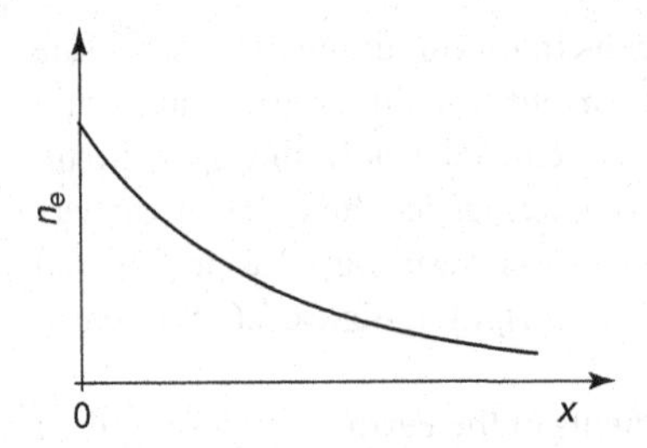

Figure 5.7 Distribution of electrons injected as minority charge carriers into a p-conductor.

To see how far an electron can travel by diffusion before it vanishes by recombination, we look at a simple example, where electrons are injected with a current density j_e into a p-conductor as minority charge carriers, as in Figure 5.7. The electrons move in the x-direction; in the y- and z-directions the system is homogeneous. The charge injected with the electrons is quickly removed by a rearrangement of the many holes in the p-conductor without, however, eliminating the electrons as particles, as explained in Section 5.4.

The steady-state distribution of the additionally injected electrons in the x-direction is given by the continuity equation

$$\frac{\partial n_e}{\partial t} = G_e - R_e - \operatorname{div} j_e = 0 \tag{5.16}$$

For the particle diffusion current,

$$j_e = -D_e \frac{dn_e}{dx} \quad \text{and} \quad \operatorname{div} j_e = -D_e \frac{d^2 n_e}{dx^2}$$

In the p-conductor,

$$G_e = G_e^0 = \frac{n_e^0}{\tau_e} \quad \text{and} \quad R_e = \frac{n_e(x)}{\tau_e} = \frac{n_e^0}{\tau_e} + \frac{\Delta n_e(x)}{\tau_e}$$

It then follows from Equation 5.16 that

$$-\frac{\Delta n_e(x)}{\tau_e} + D_e \frac{d^2 \Delta n_e(x)}{dx^2} = 0 \tag{5.17}$$

The solution takes the form

$$\Delta n_e(x) = \Delta n_e(0) \exp\left(-\frac{x}{L_e}\right) \tag{5.18}$$

The characteristic length L_e is the diffusion length (here for the electrons). Substituting this result into Equation 5.17 we obtain

$$L_e = \sqrt{D_e \tau_e} \tag{5.19}$$

For electrons in pure silicon $D_e = 35\ \text{cm}^2\ \text{s}^{-1}$. For a lifetime of, e.g. $\tau_e = 10^{-6}$ s, $L_e = 60\ \mu\text{m}$. In pure silicon, diffusion lengths of a few millimeters are obtained for electrons.

The diffusion length is the mean path length for the diffusion of a charge carrier during its lifetime. In the homogeneously illuminated pn structure in Figure 5.6 only those electrons that are generated in the p-conductor at a distance not larger than their diffusion length will reach the n-conductor.

Once the electrons are within the n-conductor, in a region in which the electrons are the majority charge carriers, their recombination has no effect on the charge current. Even though they recombine in the n-conductor, recombination does not eliminate their charge. Since the charge is transported by the majority carriers, as is explained in the next section, other electrons continue to transport the charge. Recombination reduces the charge current only if it affects the minority carriers produced by the illumination.

5.4 Dielectric Relaxation

In our discussion of the diffusion of electrons in a p-conductor in the previous section, we did not consider the formation of space charge resulting from the charge of the injected electrons, and its effect on the transport of the electrons. We now show that any space charge of minority carriers is rapidly removed by a rearrangement of the majority carriers. In an otherwise completely homogeneous system with conductivity σ, a space charge $\rho_Q(x)$ is assumed to exist at the time $t = 0$. We now examine how rapidly it decays via the charge current j_Q driven by the field of the space charge. We assume that the majority carriers dominate the conductivity and that their rearrangement causes only negligible changes of their concentration, so that diffusion is not important.

The continuity equation for the charge takes the form

$$\frac{\partial \rho_Q}{\partial t} = -\mathrm{div}\, j_Q \tag{5.20}$$

Because charge is strictly conserved, there are no rates of production and annihilation. One of Maxwell's equations relates the space charge density ρ_Q to the electric field strength E

$$\mathrm{div}\, D = \epsilon\epsilon_0 \mathrm{div}\, E = \rho_Q \tag{5.21}$$

The electric field produces a charge current $j_Q = \sigma E$, from which we find

$$\mathrm{div}\, j_Q = \sigma \mathrm{div}\, E + E\, \mathrm{grad}\, \sigma \tag{5.22}$$

Because of the assumed homogeneity, $\mathrm{grad}\, \sigma = 0$. Substituting Equation 5.22 into Equation 5.20 and replacing $\mathrm{div}\, E$ from Equation 5.21,

$$\frac{\partial \rho_Q}{\partial t} = -\sigma \mathrm{div}\, E = -\frac{\sigma}{\epsilon\epsilon_0}\rho_Q \tag{5.23}$$

with the solution

$$\rho_Q(t) = \rho_Q(0) \exp\left(\frac{-t}{\epsilon\epsilon_0/\sigma}\right) \tag{5.24}$$

$\epsilon\epsilon_0/\sigma$ is known as the *dielectric relaxation time*. If it is small compared with the lifetime of the minority carriers, their motion is essentially unaffected by the initial space charge. In this example, the diffusion of electrons into the p-conductor of Figure 5.6 pulls compensating charge by dielectric relaxation of the majority carriers (holes) through the contacts into the p-conductor.

In p-silicon, with a doping of 10^{17} cm^{-3}, a hole mobility of $b_h = 480$ cm^2 V^{-1} s^{-1} and $\epsilon = 12$, the dielectric relaxation time is $\epsilon\epsilon_0/\sigma = 1.4 \times 10^{-13}$ s, and, consequently, several orders of magnitude smaller than the lifetimes.

5.5
Ambipolar Diffusion

For the treatment of diffusion in Section 5.3 we have assumed that the diffusing particles were minority carriers with a concentration much smaller than that of the majority carriers. As a result, the space charge accompanying the injected minority carriers had been completely eliminated by dielectric relaxation of the majority carriers.

Now we discuss the case of electrons and holes produced by light absorption close to the surface, both in large concentrations compared with the concentrations in the dark state. Owing to their concentration gradients they diffuse from the surface into the interior. We assume that the electrons have the larger mobility and larger diffusion coefficient, and, therefore, move faster. Electrons and holes therefore partly separate. This leads to the formation of a positive space charge near the surface caused by the holes left behind and a negative space charge due to the electrons in the interior. The electric field caused by this charge distribution is directed so as to compensate for the different mobilities of electrons and holes. The electrons are slowed down and the holes are accelerated. This coupled motion for a strong, nonhomogeneous excitation is referred to as *ambipolar diffusion*. Since this involves the motion of electrons and holes with the same velocity in the same direction, there is no charge current associated with ambipolar diffusion.

The particle currents of the electrons and the holes are

$$j_e = -D_e \operatorname{grad} n_e - \frac{\sigma_e}{e} E \tag{5.25}$$

$$j_h = -D_h \operatorname{grad} n_h + \frac{\sigma_h}{e} E$$

Since the charge current vanishes,

$$j_Q = eD_e \operatorname{grad} n_e - eD_h \operatorname{grad} n_h + (\sigma_e + \sigma_h)E = 0 \tag{5.26}$$

we find that the electric field associated with ambipolar diffusion is given by

$$E = \frac{e}{\sigma_e + \sigma_h}\left(D_h \operatorname{grad} n_h - D_e \operatorname{grad} n_e\right) \tag{5.27}$$

It is because of this field that the electrons and holes move with the same velocity, so that $j_e = j_h$. Since very small differences in the concentrations already produce large space charges and high field strengths, the distributions of electrons and holes cannot be very different and grad $n_e \approx$ grad $n_h =$ grad n. The resulting approximation for the field in Equation 5.27 substituted into Equation 5.25 gives the particle currents in the field E

$$j_e = j_h = -\frac{D_e\sigma_h + D_h\sigma_e}{\sigma_e + \sigma_h}\,\mathrm{grad}\,n \tag{5.28}$$

Since this has the form of the diffusion current equation, we can define

$$D_{amb} = \frac{D_e\sigma_h + D_h\sigma_e}{\sigma_e + \sigma_h} \tag{5.29}$$

as the ambipolar diffusion coefficient.

5.6 Dember Effect

To conclude this chapter, we present an example to show how important it is to distinguish between electrical potential differences and electrochemical potential differences.

We calculate the electrical potential difference resulting from the electric field of ambipolar diffusion. This means that the same assumptions of strong, inhomogeneous excitation apply, as a result of which the minority carrier concentration is greater than the majority carrier concentration in the dark. With the Einstein relation $D_i = b_i kT/e$, we express the diffusion coefficients in Equation 5.27 by the mobilities, apply the approximation that electrons and holes have the same spatial distribution (grad $n_e \approx$ grad n_h), multiply both numerator and denominator in Equation 5.27 by $e(b_e + b_h)$, and find that

$$E = \frac{kT}{e}\frac{b_h - b_e}{b_e + b_h}\frac{\mathrm{grad}(\sigma_e + \sigma_h)}{\sigma_e + \sigma_h} = \frac{kT}{e}\frac{b_h - b_e}{b_e + b_h}\,\mathrm{grad}\ln(\sigma_e + \sigma_h) \tag{5.30}$$

Using $E = -\mathrm{grad}\varphi$, following integration, we find the electrical potential difference between the surface ($x = 0$) and the interior of the semiconductor ($x = \infty$) resulting from inhomogeneous exposure to light. This is the Dember voltage,

$$\Delta\varphi_D = \varphi(0) - \varphi(\infty) = \frac{kT}{e}\frac{b_e - b_h}{b_e + b_h}\ln\left(\frac{\sigma_e(0) + \sigma_h(0)}{\sigma_e(\infty) + \sigma_h(\infty)}\right) \tag{5.31}$$

For the concrete case in which electrons and holes diffuse from the surface of a semiconductor, where they are produced, into the interior, with the electrons being more mobile ($b_e > b_h$), the surface becomes positively charged and assumes a potential, which is more positive by $\Delta\varphi_D$ than in the interior.

To correctly understand the meaning of the Dember effect, we examine the case in which the electrons are mobile, while the holes are immobile ($b_h = 0$, $\sigma_h = 0$). For this case, the Dember voltage reaches its maximum possible value,

$$\Delta\varphi_D = \frac{kT}{e} \ln \frac{n_e(0)}{n_e(\infty)} \tag{5.32}$$

For this special case, however, we can easily see that this electrical potential difference cannot be measured as a steady-state voltage. For immobile holes, the electron particle current must vanish together with the charge current under the condition of open circuit during a voltage measurement. From Equation 5.13 the electrochemical potential η_e for the electrons must have the same value everywhere when the electron current is zero. Thus, no measurable voltage is found as the difference of the electrochemical potentials between the left and right sides. Figure 5.8 shows the distribution of the potentials. The electrical potential difference in Equation 5.32 can be read directly from the separation between the conduction band edge ε_C and the Fermi energy ε_{FC} in Figure 5.8. Its value in Equation 5.32 is thus based on a constant Fermi energy with the consequence that no voltage is measurable.

A Dember potential difference (we avoid the term voltage, which we reserve for measurable voltages) always arises from an inhomogeneous generation of electrons and holes when these have different mobilities. As expected, the Dember potential difference vanishes when the electrons and holes have the same mobilities and is a maximum when one of the carrier types is immobile.

The Dember effect becomes less important in a semiconductor with metal contacts. A concentration gradient, such as assumed in our example, can only develop at a surface where the recombination rate is not especially high. Taking into account that voltages are measured between metal contacts at which, due to the large recombination rate at this interface, the electron and hole densities remain unchanged even when exposed to light, we then find the potential distribution for immobile holes between two identical metal contacts shown in Figure 5.9. Here, the production of electron–hole pairs has been extended to a small region close to the surface in order to be able to see any changes in the concentration. Once again, we see electrical potential differences between the surface and the interior. Because of the unchanged concentrations at the metal contacts, however, the Dember effect

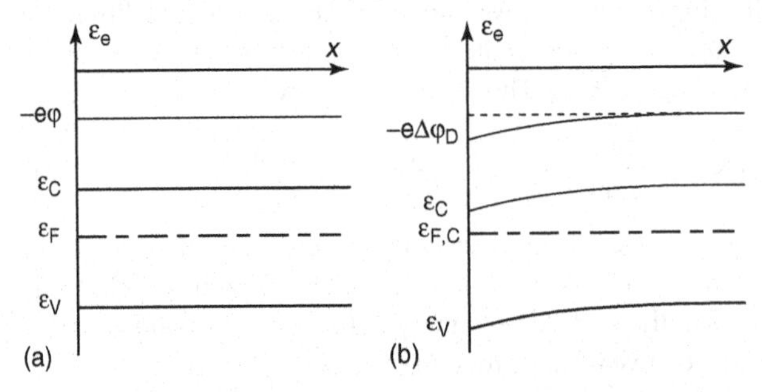

Figure 5.8 Distribution of potentials in the (a) dark state and (b) on illumination of the free left surface of a semiconductor in which only electrons are mobile.

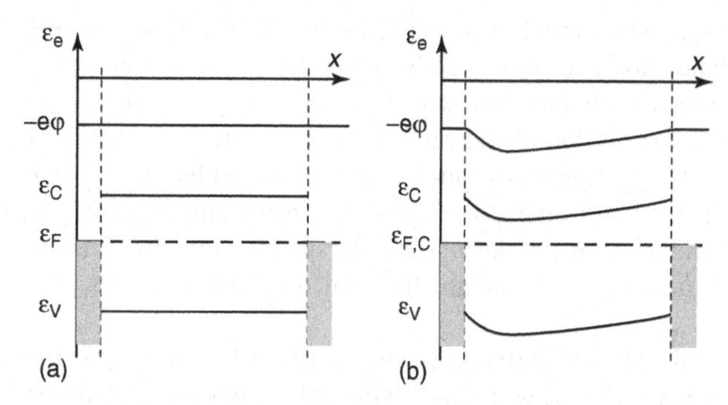

Figure 5.9 Distribution of potentials in a semiconductor with metal contacts in which only electrons are mobile, (a) in the dark state and (b) with the left surface exposed to light.

does not produce either an electrical or an electrochemical steady-state potential difference between the metal contacts when only one charge carrier type is mobile.

From this discussion of the Dember effect, we conclude that no electrochemical potential difference arises as a result of inhomogeneous illumination when only one charge carrier type is mobile and when they are distributed according to Fermi statistics over the available states. This restriction to the Fermi distribution is necessary in order to exclude the photoemission of electrons from semiconductors and metals into vacuum (or into other media), which does result in measurable voltages.

A voltage deriving from charge carriers behaving according to the Fermi distribution requires at least two different mobile carrier types. This is true, in general. For solar cells, these are electrons and holes; for a battery, these are electrons and ions.

For the preceding discussion, it was important to distinguish between electrical and electrochemical potential differences and we emphasize the essential differences once again.

Electrochemical potential differences are measurable as *steady-state* voltages and are associated with currents. The driving forces for the currents are the gradients of the electrochemical potentials and the voltage drop resulting from a charge current through a resistor in a steady state is an electrochemical potential difference.

By contrast, electrical potential differences, as seen already from their relation to the charge density in Poisson's equation, are related to the charge distribution. Temporal changes to the charge distribution can produce measurable voltages, which are, however, *nonstationary*. As an example, we can take the Dember effect for an illuminated, free surface of a semiconductor without contact, connected on the unilluminated rear side to one terminal of a voltmeter. The second terminal of the voltmeter is connected to a transparent, conductive layer in close proximity to the illuminated surface. In this arrangement, the voltmeter is capacitively coupled to the semiconductor surface. If the surface of the semiconductor becomes positively charged relative to the interior during illumination, the transparent electrode connected by the voltmeter to the rear contact of the semiconductor becomes negatively charged. The charge transport through the voltmeter then produces a voltage called *surface photovoltage*. It depends on the series connection of the capacitance between the transparent electrode and the surface, and the capacitance between the centers of positive and negative charge in the semiconductor. Here, it is important that the voltage shown by the voltmeter is *nonstationary* and decays after the RC time of the voltmeter measurement circuit although the semiconductor surface is still illuminated.

Just as electrochemical potential differences are distributed according to the values of series-connected resistances through which the same current flows, electrical potential differences are distributed according to the (reciprocal) values of series-connected capacitances, which carry the same charge.

5.7 Mathematical Description

We have now gathered all the equations, which, together with the appropriate boundary conditions, allow us to solve all problems of semiconductor devices. These equations are as follows:

- For the charge current

$$j_Q = -e j_e + e j_h \tag{5.33}$$

- For the particle currents

$$j_e = -\frac{\sigma_e}{e^2} \operatorname{grad} \eta_e \tag{5.34}$$

$$j_h = -\frac{\sigma_h}{e^2} \operatorname{grad} \eta_h \tag{5.35}$$

- For the carrier concentrations

$$\frac{\partial n_e}{\partial t} = G_e - R_e - \operatorname{div} j_e \tag{5.36}$$

$$\frac{\partial n_h}{\partial t} = G_h - R_h - \operatorname{div} j_h \tag{5.37}$$

- For the electrical potential

$$\nabla^2 \varphi = -\frac{\rho_Q}{\epsilon \epsilon_0} \tag{5.38}$$

This system of equations must be solved for every point in the solar cell subject to the boundary conditions. This is a complicated numerical procedure. In the discussion of the solar cell, we will not proceed in this direction, since it is rather abstract and conceals the physics behind the solar cell, rather than elucidating it.

5.8 Problems

5.1 Derive the so-called Einstein relation between diffusion coefficient and mobility for particles with concentration n_i, charge $z_i e$ and mobility b_i from Ficks law of diffusion, $j_{Q,i} = -z_i e D_i \operatorname{grad} n_i$ and from the general equation $j_{Q,i} = -(\sigma_i/z_i e) \operatorname{grad} \eta_i$.

5.2 A charge current density $j_Q = e j_h = 150\ \text{mA cm}^{-2}$ is flowing in a p-type silicon sample ($n_A = 10^{16}\ \text{cm}^{-3}$, assume full ionization of the acceptors) having a thickness of $d = 200\ \mu\text{m}$. The direction of the current shall be in the positive x direction from $x = 0$ to $x = d$.

(a) How large is the (hole) conductivity for a hole mobility of $b_h = 300\ \text{cm}^2\ \text{V}^{-1}\ \text{s}^{-1}$?

(b) The concentration of the holes shall be homogeneous. Determine the electrical potential difference between the two contacts of the semiconductor.

(c) If the current is caused by diffusion, how does the concentration profile of the holes look like if their total number remains unchanged?

Compare the gradients of the electrochemical potential η_h for both cases.

5.3 Assume a mobility of electrons in a semiconductor of $b_e(T = 300\ \text{K}) = 300\ \text{cm}^2\ \text{V}^{-1}\ \text{s}^{-1}$ depending on temperature as $T^{-3/2}$.
An electric field of $E = 100\ \text{V cm}^{-1}$ is applied. At which temperature are the drift velocity $\langle v_{\text{drift,e}} \rangle$ and the thermal velocity $\langle v_{\text{th,e}} \rangle$ related as $\langle v_{\text{drift,e}} \rangle = 0.001 \langle v_{\text{th,e}} \rangle$? Assume $m_e^* = m_e$.

5.4 Intrinsic carrier densities of electrons and holes can be achieved in a doped semiconductor by compensation. Compensation means that the effect of the donors with regard to the carrier concentration in the valence and conduction band is compensated by a proper amount of acceptors. Has this material the same conductivity as an impurity free, intrinsic semiconductor?

6
Basic Structure of Solar Cells

To properly understand the requirements for a solar cell, namely, to generate current and voltage, we first turn to a similar problem that may be easier to understand, and which demonstrates that the working of a solar cell has more of a chemical than an electrical nature.

6.1
A Chemical Solar Cell

Figure 6.1 illustrates a hypothetical chemical solar cell in which water (H_2O) is decomposed into hydrogen (H_2) and oxygen (O_2) by the absorption of high-energy photons. As a result, the partial pressures of hydrogen and oxygen in the cell rise above their equilibrium values. If no hydrogen and oxygen are removed, that is, in the open-circuit state under steady-state conditions, the partial pressures reach values at which the reverse reaction, that is, the recombination of hydrogen and oxygen to water, occurs at the same rate as the decomposition reaction. Clearly, the partial pressures decrease with increasing probability of the reverse reaction. With a catalyst, such as a large platinum surface, the reverse reaction would have such a high probability that, even with the additional production of hydrogen and oxygen by decomposition, the partial pressures remain practically at the equilibrium values.

As with the electrons and holes in a solar cell, our aim is to extract the gases separately along with their chemical energy. How do we remove the gases separately from the mixture in the reaction cell? This requires membranes that selectively transmit only one of the gases. A semipermeable membrane on the left in Figure 6.1 transmits the hydrogen and blocks the oxygen, while another membrane on the right is permeable for oxygen and blocks the hydrogen. If the partial pressure, and, therefore, the chemical potential, of one gas type is greater in the reaction cell than outside, this gas flows through its membrane and out of the cell. For a membrane with good permeability, only a slight difference in the partial pressures or the chemical potentials between the inside and the outside is sufficient for the transport of the gases. Similarly, a gradient of the partial pressure or the chemical potential is required inside the cell as the driving force for the transport of the molecules

Physics of Solar Cells: From Basic Principles to Advanced Concepts. Peter Würfel

ISBN: 978-3-527-40857-3

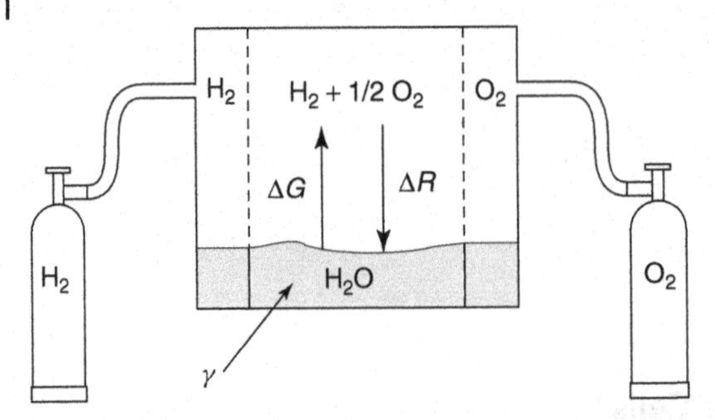

Figure 6.1 Hypothetical chemical solar cell in which water is decomposed into hydrogen and oxygen by the absorption of photons. Hydrogen and oxygen can be separately removed through membranes that selectively pass hydrogen on the left and oxygen on the right.

toward the appropriate membranes. How is this gradient established although the decomposition rate generating the gas molecules is homogeneous? The gradients are an automatic consequence of allowing the molecules to pass through the appropriate membrane, which reduces the partial pressure or chemical potential at the membrane and sets up the gradients in the right direction. The chemical energy obtained from the decomposition of the water is greatest in the open-circuit state, when the gases are not removed and recombine. But in the absence of gas flow, no chemical energy is removed. On the other hand, removing the gases decreases their partial pressures. For the other limiting case, when all of the gases produced are removed and the recombination rate is at its equilibrium value, the partial pressures and chemical potentials are also at their equilibrium values and consequently no chemical energy is supplied with the gas flow. To obtain the maximum chemical energy current, we must therefore accept a certain level of recombination. In the following, we analyze the performance of this chemical solar cell quantitatively.

From the minimum of the free energy for the equilibrium of all components in the chemical cell at constant temperature and volume and without illumination

$$dF = \mu_{H_2}\, dN_{H_2} + \mu_{O_2}\, dN_{O_2} + \mu_{H_2O}\, dN_{H_2O} = 0 \tag{6.1}$$

From the chemical reaction

$$H_2 + \tfrac{1}{2} O_2 \rightleftharpoons H_2O$$

changes in the number of the molecules are related by

$$dN_{H_2} = 2\, dN_{O_2} = -dN_{H_2O}$$

and Equation 6.1 becomes

$$\mu_{H_2} + \tfrac{1}{2}\mu_{O_2} = \mu_{H_2O} \tag{6.2}$$

which, as we have seen before, could have been read directly from the reaction equation. Equation 6.2 is valid for chemical equilibrium. When the sum of the chemical potentials of the gases is larger than the chemical potential of the water, there is a net production of water from the gases. When the sum of the chemical potentials of the gases is smaller than the chemical potential of the water, there is a net production of the gases from the dissociation of water. The chemical potentials of the gases depend on their concentrations in the same way as for electrons and holes, whereas the chemical potential of the water molecules in the water does not change when the amount of water changes, since the concentration of water molecules stays constant.

$$\mu_{gas} = \mu_0 + kT \ln\left(\frac{p_{gas}}{p_0}\right) \tag{6.3}$$

where the proportionality between concentration and partial pressure p was used. The standard pressure p_0 is usually chosen as 1 bar under standard conditions. With the relation in Equation 6.3, a relation for the partial pressures of hydrogen and oxygen in chemical equilibrium with liquid water at 300 K, which is the equivalent of the relation between the concentrations of electrons and holes in an unilluminated semiconductor, $n_e^0 n_h^0 = n_i^2$, is found from Equation 6.2.

$$kT \ln\left(\frac{p_{H_2}^0 \sqrt{p_{O_2}^0}}{p_0^{3/2}}\right) = \mu_{0,H_2O} - \mu_{0,H_2} - \frac{1}{2}\mu_{0,O_2} = -2.46 \text{ eV} \tag{6.4}$$

The value on the right-hand side of Equation 6.4 is found in books on physical chemistry. As for the equilibrium concentrations of electrons or holes by doping, any one of the partial pressures can be chosen. When the partial pressure of oxygen is $p_{O_2}^0 = 0.2$ bar as in air, the equilibrium partial pressure of hydrogen is only $p_{H_2}^0 = 1.8 \times 10^{-41}$ bar. If the hydrogen partial pressure is reduced below this value, the recombination rate of the gases is smaller than the thermal rate of splitting of water molecules, until the equilibrium partial pressure is reestablished. Since 1 H_2 molecule reacts with $\frac{1}{2}O_2$ molecule to give 1 molecule of H_2O, the recombination rate of hydrogen is

$$R_{H_2} \sim p_{H_2} \sqrt{p_{O_2}}$$

This recombination rate for nonequilibrium partial pressures can be expressed by the equilibrium recombination rate

$$R_{H_2} = R_{H_2}^0 \frac{p_{H_2} \sqrt{p_{O_2}}}{p_{H_2}^0 \sqrt{p_{O_2}^0}} \tag{6.5}$$

When the generation rate is increased from the equilibrium generation rate $G_{H_2}^0$ by ΔG_{H_2} by illumination, hydrogen may flow out from the reaction cell through its membrane. The hydrogen current, arbitrarily counted negative if it is flowing out of the cell, is

$$j_{H_2} = 2j_{O_2} = R_{H_2} - G_{H_2}$$

$$j_{H_2} = 2j_{O_2} = R^0_{H_2}\frac{p_{H_2}\sqrt{p_{O_2}}}{p^0_{H_2}\sqrt{p^0_{O_2}}} - G^0_{H_2} - \Delta G_{H_2}$$

$$j_{H_2} = 2j_{O_2} = R^0_{H_2}\left(\frac{p_{H_2}\sqrt{p_{O_2}}}{p^0_{H_2}\sqrt{p^0_{O_2}}} - 1\right) - \Delta G_{H_2} \tag{6.6}$$

The interpretation is simple. Hydrogen and oxygen flow into the cell in a steady state only if more molecules in the cell disappear by recombination than are generated by water splitting. On the other hand, molecules flow out of the cell only if more are generated in the cell than recombine. If we assume that the gradients of the partial pressures required to drive the currents are small, the partial pressures in the cell and in the gas bottles in Figure 6.1 are approximately the same, as is also shown in Figure 6.2. As a result, the currents of hydrogen and oxygen are limited only by the rates of their chemical reaction and not by transport resistances.

The hydrogen current flowing between the hydrogen bottle and the cell is shown as a function of the hydrogen pressure in the hydrogen bottle, while the oxygen pressure is kept constant (Figure 6.3). The current changes with the hydrogen pressure only because of the recombination rate. The generation rate, in the dark and under illumination does not depend on the hydrogen pressure. At zero hydrogen pressure in the cell and outside, there is no recombination and all molecules generated flow out of the cell. The open-circuit pressure at zero current, when the recombination rate equals the generation rate, is the equilibrium pressure in the dark and a larger pressure with illumination. For still larger pressures, more molecules disappear by recombination than are generated. They are replenished by a positive current flowing into the cell.

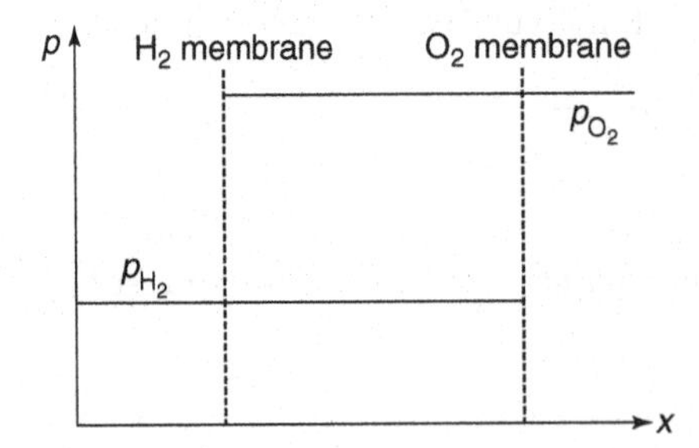

Figure 6.2 The conductivities for hydrogen in the hydrogen membrane on the left and for oxygen in the oxygen membrane on the right, are assumed to be large. The gradients of the partial pressures required to drive the currents are therefore small and do not show up in the spatial distribution of the pressures.

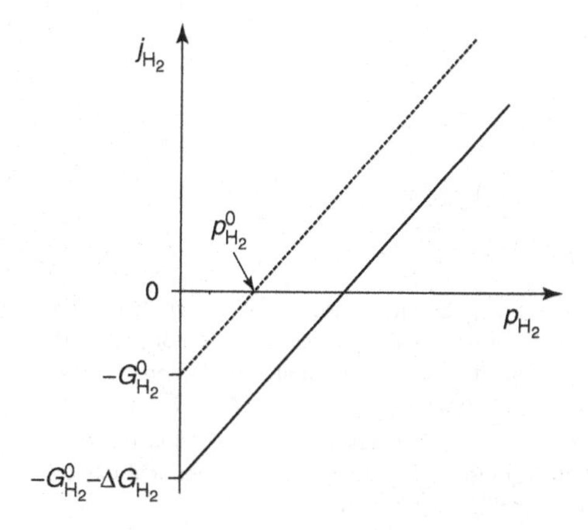

Figure 6.3 Current of hydrogen, positive if flowing from the hydrogen bottle in Figure 6.1 into the cell as a function of the hydrogen partial pressure, without illumination (broken line) and with additional generation ΔG_{H_2} by illumination (solid line). The oxygen pressure is kept constant.

We could also plot the hydrogen current as a function of the chemical potential of the hydrogen while keeping the chemical potential of the oxygen constant. This plot has the advantage that the chemical energy exchanged between the gas bottles and the cell shows up as a rectangle as for the electron–hole current in Figure 4.4. With Equations 6.2 and 6.3, the recombination rate can be written as

$$R_{H_2} = R^0_{H_2} \exp\left(\frac{\mu_{H_2} + 1/2\,\mu_{O_2} - \mu_{H_2O}}{kT}\right) \tag{6.7}$$

and the dependence of the hydrogen currents on the chemical potentials is

$$j_{H_2} = 2j_{O_2} = R^0_{H_2}\left[\exp\left(\frac{\mu_{H_2} + 1/2\,\mu_{O_2} - \mu_{H_2O}}{kT}\right) - 1\right] - \Delta G_{H_2} \tag{6.8}$$

which is shown in Figure 6.4.

6.2 Basic Mechanisms in Solar Cells

We now turn to a semiconductor solar cell in which, instead of hydrogen and oxygen, additional electrons and holes are generated by illumination. This is the only difference from the chemical cell in the last section. The thorough discussion of the chemical cell was intended to show the parallels and to emphasize the importance of the chemical reaction between electrons and holes in semiconductors.

What do we learn from the chemical solar cell? The most important point is that semipermeable membranes are required on both sides of the absorber, such that electrons can only flow out to the left and holes to the right, forcing them to generate a charge current flowing from left to right in the cell.

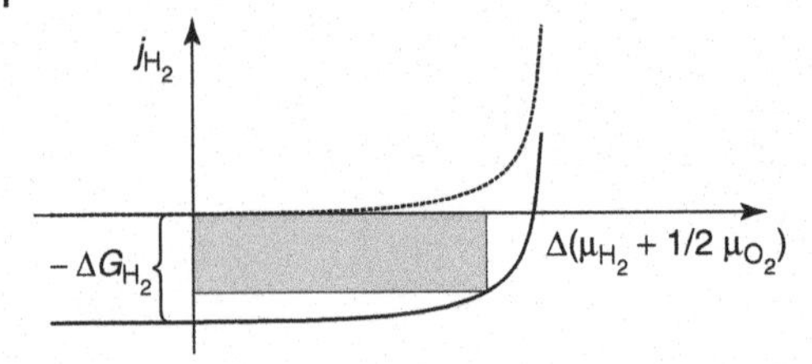

Figure 6.4 Current of hydrogen, positive if flowing from the hydrogen bottle in Figure 6.1 into the cell as a function of the deviation $\Delta(\mu_{H_2} + 1/2\mu_{O_2})$ of the chemical potentials of hydrogen and oxygen from their equilibrium values, without illumination (broken line) and with additional generation ΔG_{H_2} by illumination (solid line). The chemical potential of water μ_{H_2O} is not changed by illumination. A smaller and more realistic equilibrium generation rate $G^0_{H_2}$ than in Figure 6.3 is assumed. The shaded rectangle is the largest current of chemical energy delivered by the cell.

A membrane that is permeable for electrons, and which blocks the holes, would be a material that has a large conductivity for electrons and a small conductivity for holes. We already know that an n-type semiconductor has this property. However, different from the hydrogen membrane, which has a very small mobility for oxygen, the hole conductivity in an n-type semiconductor is small because of a small concentration of holes. To function properly as an electron membrane, we have to make sure that holes are not injected from the absorber. A hole membrane is, of course, a p-type semiconductor, into which electrons must not be injected. Figure 6.5 shows a solar cell structure which has the required properties. The injection of holes into the electron membrane on the left is prevented by giving it a larger band gap, resulting in an energy barrier in the valence band for the holes. In the same way, a larger band gap of the hole membrane on the right, combined with a smaller electron affinity χ_e causes an energy barrier in the conduction band for the electrons. Owing to their larger band gap, the membranes transmit almost all of

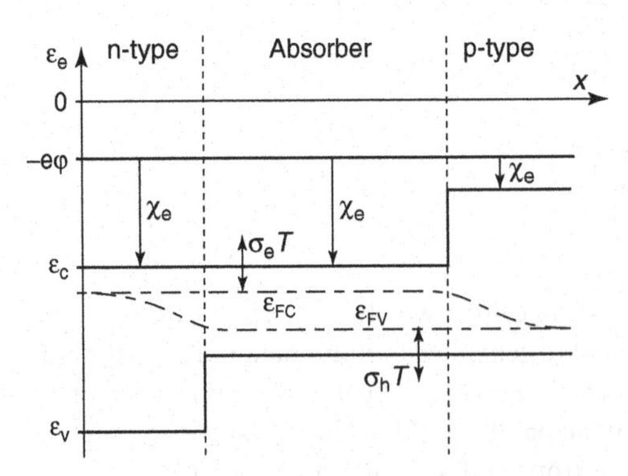

Figure 6.5 A n-type electron membrane on the left allows electrons generated in the absorber by illumination, to flow to the left, while blocking the holes. A p-type hole membrane on the right allows holes to flow to the right, blocking the electrons. Electrons are driven by an invisibly small gradient of $-\varepsilon_{FC}$, holes are driven by an invisibly small gradient of ε_{FV}.

the photons to be absorbed in the absorber. As for the partial pressure distribution in Figure 6.3, transport resistances for electrons and holes are assumed to be small (which is realistic) and hence, gradients of the Fermi energies required to drive the electrons to the left and the holes to the right are negligibly small. Metal contacts are assumed to make contact with the membranes, which forces the Fermi energies to join at the surface. The large gradient of ε_{FV} in the electron membrane on the left does not lead to an appreciable hole current to the left because of the very small hole concentration in the electron membrane, obvious from the large distance of ε_{FV} from the valence band. Accordingly, the same holds for the electrons in the hole membrane.

For the conversion of chemical energy into electrical energy in a solar cell we have found a structure in which the electrons and holes flow outward through different contacts. This structure is virtually identical with the structure for the removal of hydrogen and oxygen from the chemical reaction cell. In the following, we discuss different arrangements that fulfill the same requirements with different degrees of perfection. Special emphasis is given to the pn-junction because of its technical importance and its model character for other devices, although it is not quite as ideal for a solar cell as the heterostructure in Figure 6.5.

6.3 Dye Solar Cell

A very good example of the solar cell structure required, with membranes for electrons on one side and for holes on the other side, is the electrochemical dye solar cell, shown in Figure 6.6 [8]. The "semiconductor" in which the absorption of photons produces electron–hole pairs is a dye layer. Since the electrons and holes in the dye layer have very small mobilities, this layer must be very thin for the charge carriers to reach the membranes within their lifetimes.

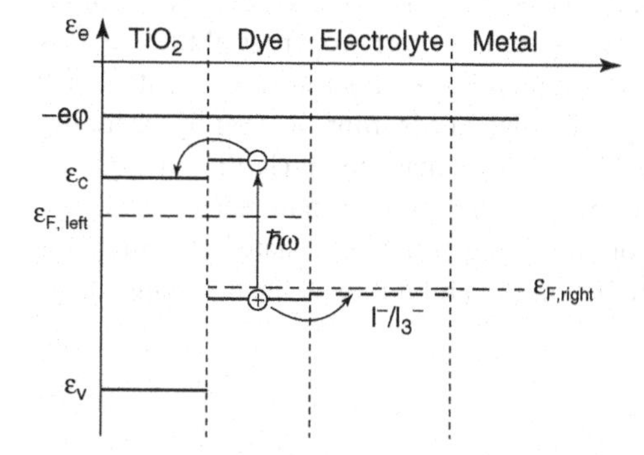

Figure 6.6 Dye solar cell in which the electron–hole pairs are produced in a ruthenium bipyridine dye. The electrons flow outward toward the left through the n-conductor TiO_2 and the holes toward the right through the triiodide ions with which the acetonitrile electrolyte is doped.

The dye is applied as a monomolecular layer to a good electron conductor in the form of TiO_2. As Figure 6.6 illustrates, the electrons in the dye reach the conduction band of the TiO_2 without difficulty. However, owing to the large band gap of more than 3 eV of TiO_2, the holes present in the dye encounter a high barrier for the transition to the valence band of the TiO_2.

On the other hand, the thin, only monomolecular dye layer required for efficient charge transfer has the disadvantage that the absorption of photons in one layer is very poor, since their penetration depth $1/\alpha$ into the dye is much larger than the thickness of the dye layer. In order to compensate for this disadvantage, the TiO_2 layer is composed of particles, only a few nanometers in size, in a porous structure. All TiO_2 particles are coated with the dye on their free surfaces, so that complete absorption of the photons is achieved with the many dye layers encountered by the photons.

The porous structure, however, greatly complicates the contact between the dye and a p-type hole membrane, through which the holes can flow outwards. This problem is solved with the use of an electrolyte that penetrates into all the pores. The iodide ions of the redox system (I^-/I_3^-) provide for charge transport. The energy of an electron in I^- differs only slightly from the energy of an electron in the ground state of the dye, so that the flow of holes from the dye to the electrolyte is unproblematic. The flow of the excited electrons from the dye into the electrolyte is, however, prevented since the electrolyte has no states at the energy of the excited electrons.

At first glance, this electrochemical cell would appear almost ideally to fulfill the requirements for the selective transport of electrons toward the left in the TiO_2 and of holes toward the right in the electrolyte. Besides direct recombination via the direct transition of an electron from its excited state in the dye to its ground state, there is also recombination via an indirect path, which we can regard as an internal shunt. Excited electrons that flow to the conduction band of the TiO_2 can also reach the ground state of the dye via surface states of the TiO_2 particles, either directly or, more probably, after a transition to the redox system. This back reaction of the electrons is facilitated by the large surface area of the TiO_2, which is, however, required in order to combine sufficient absorption with the poor transport properties of the dye. It is therefore not certain whether the dye solar cell represents a solar cell suitable for practical application. The absorption of the dye needs improvement by extending it over a greater spectral region. Problems that remain to be clarified include its stability over a period of 20 years with respect to decomposition reactions and the possibility of the electrolyte leaking out or drying out. A solid-state hole conductor would be preferable, but a way has to be found to fill it into all the tiny pores.

6.4
The pn-Junction

A good, although not ideal, realization of the structure of a solar cell as discussed in the previous section can be found in commercially available solar cells made of crystalline silicon. A p-region about 300 μm thick, moderately doped with an acceptor concentration of $n_A = (10^{15} - 10^{16})\ cm^{-3}$, as the absorber is sandwiched between a less than 1 μm thick, highly doped n-layer on the illuminated side, as the electron membrane and another thin, highly doped p-layer on the rear side, as the hole membrane. All of this consists of crystalline silicon. The pn-junction, formed by the electron membrane and the absorber, near which most electrons and holes are generated, is especially important for the solar cell and other devices, and is therefore the subject of a more detailed discussion.

6.4.1
Electrochemical Equilibrium of Electrons in a pn-Junction in the Dark

In thermal equilibrium with the environment, including the 300 K background radiation, no current may flow in the pn-junction if there is no external energy source. This means that

1. $j_Q = 0$

   and as a result of chemical equilibrium with the 300 K radiation,

2. $\eta_e + \eta_h = \mu_\gamma = 0$

From Equation 5.13, j_Q is given by

$$j_Q = \frac{\sigma_e}{e}\text{grad}\,\eta_e - \frac{\sigma_h}{e}\text{grad}\,\eta_h = 0$$

From (2) above, it follows that grad $\eta_e = -$grad η_h and therefore

$$j_Q = \frac{\sigma_e + \sigma_h}{e}\text{grad}\,\eta_e = 0$$

Since $\sigma_e + \sigma_h \neq 0$, grad $\eta_e = 0$.

This tells us that, in the dark, the electrochemical potential η_e of the electrons (as well as the electrochemical potential η_h of the holes) has the same value everywhere in the pn-junction. This is the implication of electrochemical equilibrium between electrons in the n-region and electrons in the p-region. Indicating values in the p-region far away from the pn-junction by a superscript p and in the n-region far away from the pn-junction by a superscript n, we have

$$\eta_e^p = \mu_{e,0}^p + kT\ln\frac{n_e^p}{N_C} - e\varphi^p = \eta_e^n = \mu_{e,0}^n + kT\ln\frac{n_e^n}{N_C} - e\varphi^n \tag{6.9}$$

Since only a small fraction of the semiconductor atoms is replaced by doping atoms, the chemical environment for a free electron, which determines the value of $\mu_{e,0}$, remains unchanged. For $\mu_{e,0}^p = \mu_{e,0}^n$ the difference in the electrical potentials between the p-region and the n-region is, from Equation 6.9,

$$\varphi^n - \varphi^p = \frac{kT}{e}\ln\frac{n_e^n}{n_e^p}$$

Using $n_e^n = n_D$ and $n_e^p = n_i^2/n_A$, this potential difference, called the *diffusion voltage*, is

$$\varphi^{n} - \varphi^{p} = \frac{kT}{e} \ln \frac{n_D n_A}{n_i^2} \tag{6.10}$$

We can visualize this potential difference as arising in the following way. We start with the spatially separated and electrically neutral p-conductor and the n-conductor being at the same electrical potential. When making contact, the greater chemical potential of the electrons in the n-conductor (and of the holes in the p-conductor) drives a diffusion current of electrons from the n-conductor to the p-conductor and a hole diffusion current from the p-conductor to the n-conductor. This builds up a positive charge in the n-conductor and a negative charge in the p-conductor. The diffusion currents continue to flow until an electrical potential difference $\varphi^n - \varphi^p$ is established for which $\eta_e^p = \eta_e^n$, so that grad $\eta_e = 0$ and grad $\eta_h = 0$. The driving force then no longer exists, and the particle currents, and, with them, the charge current can no longer flow.

For electrochemical equilibrium, in which the gradient of the electrical energy is compensated by the gradient of the chemical energy, the electrons and holes do not experience any forces. Yet, they are mobile in random Brownian motion.

6.4.2
Potential Distribution across a pn-Junction

The potential difference $\varphi^n - \varphi^p$ is the result of the electrochemical equilibrium of the electrons in the n- and p-regions. The distribution of this potential difference over the n- and p-conductors follows from the relationship with the distribution of the charge density ρ_Q.

From the Maxwell equation

$$\text{div } D = \rho_Q$$

with $D = \epsilon\epsilon_0 E$ and with $E = -\text{grad } \varphi$, we find Poisson's equation

$$\text{div } E = -\text{div grad } \varphi = -\nabla^2 \varphi = \frac{\rho_Q}{\epsilon\epsilon_0} \tag{6.11}$$

The interface between the n- and p-regions is assumed to extend much farther in the y- and z-directions than the width of the space charge layer to be derived. This allows a one-dimensional treatment, giving

$$\frac{d^2\varphi}{dx^2} = -\frac{\rho_Q}{\epsilon\epsilon_0} \tag{6.12}$$

In the n-region the density of the space charge is

$$\rho_Q^n(x) = e\left(n_D^+ - n_e(x)\right) = e\, n_D^+ \left(1 - \exp\left\{\frac{e\left[\varphi(x) - \varphi^n\right]}{kT}\right\}\right) \tag{6.13}$$

This relationship follows from Figure 6.7, which schematically depicts the potential distribution and the charge distribution.

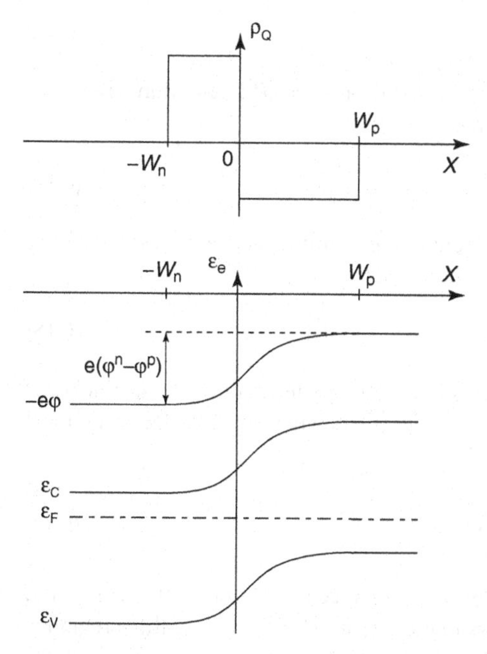

Figure 6.7 Distribution of the space charge density ρ_Q and of the electrical energy $-e\varphi$ per electron in a pn-junction.

Unfortunately, however, for this space charge density, Poisson's equation can only be solved numerically. According to Schottky, we can approximate the charge distribution by a spatially constant positive space charge density in a region up to an, as yet, undefined depth w_n on the n-side. A constant negative space charge density is assumed up to w_p in the p-region, which compensates the positive space charge in the n-region, as shown in Figure 6.7.

$$\rho_Q^n = e\,n_D^+ \approx e\,n_D \quad \text{for} \quad -w_n < x \le 0$$

$$\rho_Q^p = -e\,n_A^- \approx -e\,n_A \quad \text{for} \quad 0 \le x < w_p$$

The sum of the charges $Q_n = en_D w_n$ in the n-region and $Q_p = -en_A w_p$ in the p-region is $Q_n + Q_p = 0$, and the electric field strength differs from zero only between $-w_n$ and w_p. From the sum of the charges

$$w_p = \frac{n_D}{n_A}\, w_n \tag{6.14}$$

and the entire thickness of the space charge layer is

$$w = w_n + w_p = \left(1 + \frac{n_D}{n_A}\right) w_n \tag{6.15}$$

For the boundary conditions $E(-w_n) = 0$ for the electric field and $\varphi(-w_n) = \varphi^n$ for the electrical potential, integration of Poisson's equation 6.12 with $\rho_Q = \rho_Q^n$ in the range $-w_n < x \le 0$ of the n-region yields

$$\varphi_n(x) = -\frac{e n_D}{2\epsilon\epsilon_0}(x + w_n)^2 + \varphi^n \tag{6.16}$$

and for the boundary conditions $E(w_p) = 0$ and $\varphi(w_p) = \varphi^p$ integration with $\rho_Q = \rho_Q^p$ in the range $0 \leq x < w_p$ of the p-region yields

$$\varphi_p(x) = \frac{e n_A}{2\epsilon\epsilon_0}(x - w_p)^2 + \varphi^p \tag{6.17}$$

For the given charge distribution the potential is continuous everywhere including at $x = 0$. From $\varphi_n(0) = \varphi_p(0)$ it thus follows that

$$\varphi^n - \varphi^p = \frac{e}{2\epsilon\epsilon_0}\left(n_D w_n^2 + n_A w_p^2\right) \tag{6.18}$$

With reference to Equations 6.14 and 6.15, we can now determine the total thickness w of the space charge layer, since $\varphi^n - \varphi^p$ is known from the electrochemical equilibrium according to Equation 6.10:

$$w = \sqrt{\frac{2\epsilon\epsilon_0}{e}\frac{n_A + n_D}{n_A n_D}(\varphi^n - \varphi^p)} \tag{6.19}$$

For the asymmetrical pn-junction of a Si solar cell, with $n_D = 10^{19}$ cm^{-3} and $n_A = 10^{16}$ cm^{-3}, Equation 6.14 shows that the space charge layer in the p-region is much thicker than in the n-region. Its extension into the p-region w_p is essentially equal to the total thickness w which, for the doping levels above, has a value of $w = 0.35$ μm. Over this distance, the change in potential is $\varphi^n - \varphi^p = 0.9$ V as follows from Equation 6.10. At least for crystalline silicon, the extension of the space charge layer is therefore much smaller than the penetration depth of the photons and the diffusion lengths.

From the discussion of the recombination mechanisms in Chapter 3, we know that high doping levels favor Auger recombination and should be avoided in a solar cell. Are there minimum doping levels required for a pn-junction solar cell? The problem is that the conversion of chemical energy of the electron–hole pairs into electrical energy is incomplete, if the difference between the Fermi energies at the contacts $\varepsilon_{F,left} - \varepsilon_{F,right}$ is smaller than the difference between the Fermi energies $\varepsilon_{FC} - \varepsilon_{FV}$ in the absorber, as shown in Figure 5.6. Incomplete conversion results in the generation of entropy. A comparison of Figure 5.6 with Figure 6.5, where the conversion is complete, indicates that the entropy per electron σ_e, which increases with the distance of the Fermi energy ε_{FC} from the conduction band, increases from the interior toward the left surface in Figure 5.6. Entropy generation is obvious, since the entropy rises at the expense of the free energy of the electrons $\eta_e = \varepsilon_{FC}$. The same happens to the entropy per hole σ_h, which increases with the distance of the Fermi energy ε_{FV} from the valence band and rises toward the right surface. This entropy generation is avoided, if care is taken that the concentration of the electrons in a pn-junction is at least the same everywhere on the path of the carriers, toward their membrane.

From the above considerations, it follows that the majority carrier concentrations in the dark in regions functioning as membranes must be at least as large as the

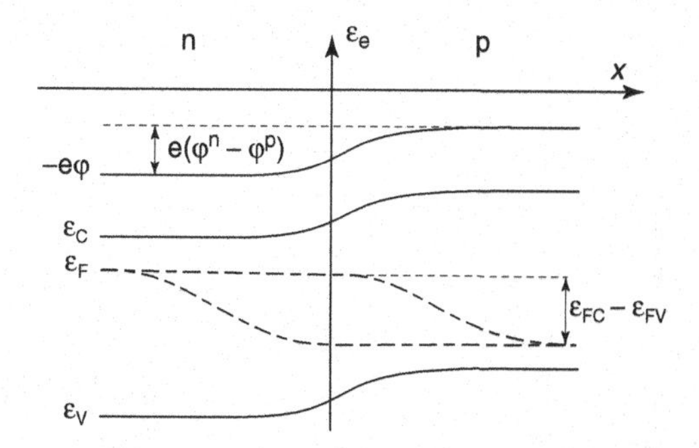

Figure 6.8 Potential distribution in an illuminated solar cell in which the diffusion voltage is $e(\varphi^n - \varphi^p) > \varepsilon_{FC} - \varepsilon_{FV}$.

additional concentrations generated by illumination. Since the dark concentrations of the majority carriers are equal to the doping concentrations, this condition is expressed as

$$n_D n_A \geq n_e^p n_h^n$$

equivalent to the condition

$$kT \ln \frac{n_D n_A}{n_i^2} = e\left(\varphi^n - \varphi^p\right) \geq \varepsilon_{FC} - \varepsilon_{FV} = kT \ln \frac{n_e^p n_h^n}{n_i^2} \tag{6.20}$$

This requires suitable doping to correctly match the potential difference $\varphi^n - \varphi^p$ in the dark to the expected chemical energy per electron–hole pair during illumination.

In Figure 6.5 we have chosen $e(\varphi^n - \varphi^p) = \varepsilon_{FC} - \varepsilon_{FV}$, so that the electrical potential difference between the p- and n-conductor membranes just vanishes during illumination. By contrast in Figure 5.6 $e(\varphi^n - \varphi^p) < \varepsilon_{FC} - \varepsilon_{FV}$. The electrical potential difference vanishes in this case as well for the assumed illumination level, but the photovoltage measured at the contacts is smaller than the difference between the Fermi energies inside the cell: hence chemical energy is wasted. Figure 6.8 gives the potential distribution for a more strongly doped cell, for which $e(\varphi^n - \varphi^p) > \varepsilon_{FC} - \varepsilon_{FV}$. The chemical energy per electron–hole pair is now completely converted into electrical energy. Because of the increasing probability of Auger recombination with increased doping, however, the doping density should not be larger than necessary.

6.4.3
Current–Voltage Characteristic of the pn-Junction

For the charge current through a pn-junction we distinguish between the forward direction, in which the electrons of the n-region and the holes of the p-region flow toward the pn-junction, and the reverse direction, in which electrons and holes flow away from the pn-junction.

In the forward direction, shown in Figure 6.9a, both the electrons coming from the n-region and the holes coming from the p-region move as minority carriers

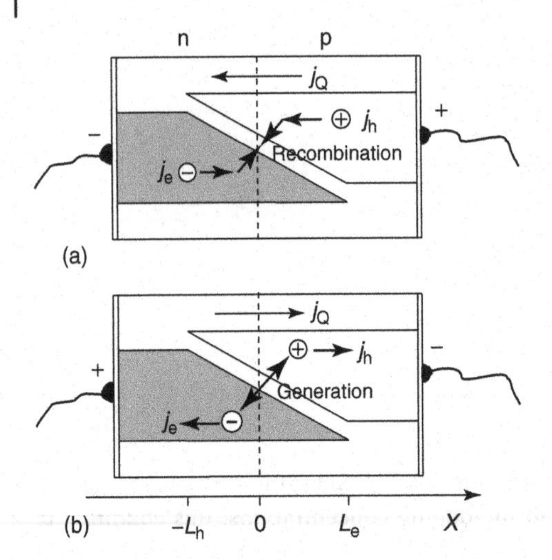

Figure 6.9 Electron and hole currents in a pn-junction. (a) For a negative polarity of the n-region with respect to the p-region, i.e. in the forward direction, electrons and holes flow toward the pn-junction, where they recombine. (b) In the reverse direction, for a positive polarity of the n-region, electrons and holes flow away from the pn-junction where they are produced.

into the oppositely doped region, where they recombine after an average path length of one diffusion length. More than a diffusion length away from the pn-junction, the minority carrier concentration is much smaller than the majority carrier concentration, both in the dark and with illumination (weak excitation), so that the charge current is carried only by the majority carriers, in the n-region by electrons and in the p-region by holes.

In the reverse direction, shown in Figure 6.9b, electrons come from the p-region and holes from the n-region. Since the charge current in the p-region is made up entirely of holes, no electrons are transported through the p-region. The electrons emerging from the p-region must have been produced there. However, only those electrons generated in the p-region that have not recombined before can reach the n-region. These electrons must have been generated at a distance not farther than a diffusion length from the n-region. For the same reason, only those holes generated at a distance not farther than a diffusion length from the p-region can reach the p-region. For both the forward and the reverse direction, the charge of the charge current is transferred from the electrons to the holes within a diffusion length on both sides of the pn-junction.

Outside an electron diffusion length L_e to the right or a hole diffusion length L_h to the left of the pn-junction, the charge current is a pure electron current in the n-region and a pure hole current in the p-region. The charge current is then given by integrating over the contributions to the hole current (alternatively, the contributions to the electron current). If the forward charge current is arbitrarily counted as positive (electrons and holes flow toward the pn-junction),

$$j_Q = -e \int_{-L_h}^{L_e} \operatorname{div} j_h \, dx \tag{6.21}$$

since $j_h = 0$ for $x < -L_h$, but $j_h = j_Q/e$ for $x > L_e$.

From the continuity equation for holes under steady state conditions

$$\frac{\partial n_h}{\partial t} = G_h - R_h - \operatorname{div} j_h = 0, \quad \operatorname{div} j_h = G_h - R_h$$

Again, we divide the generation rate into the components G_h^0 in the dark and ΔG_h from the illumination

$$G_h = G_h^0 + \Delta G_h$$

The recombination rate for radiative recombination is accordingly

$$R_h = R_h^0 \frac{n_e n_h}{n_i^2} = R_h^0 \exp\left(\frac{\eta_e + \eta_h}{kT}\right) \tag{6.22}$$

where, because of $\eta_e + \eta_h = 0$ in equilibrium with the 300 K photons and phonons,

$$G_h^0 = R_h^0$$

We can now write the charge current of Equation 6.21 as

$$j_Q = -e \int_{-L_h}^{L_e} \left\{ G_h^0 \left[1 - \exp\left(\frac{\eta_e + \eta_h}{kT}\right)\right] + \Delta G_h \right\} dx \tag{6.23}$$

In principle, the sum of the electrochemical potentials $\eta_e + \eta_h$ is a function of the position and depends on the resistances that limit the charge current.

Resistances, however, occur as two different types:

1. the familiar type, when the transport resistance is limiting the current. Then

$$j_Q = -\frac{\sigma}{e} \operatorname{grad} \eta$$

2. the less familiar type, when the current is limited by the resistance of a chemical reaction as in the reaction of hydrogen and oxygen in the chemical solar cell in Section 6.1 and of electrons and holes with photons and phonons in a semiconductor cell

$$e + h \rightleftharpoons \gamma, n\Gamma$$

In this case,

$$\operatorname{div} j = G - R$$

If the chemical reaction limits the current, as it does in the chemical solar cell, no more electrons and holes can flow away from the pn-junction than are produced there, and no more can flow toward the pn-junction than disappear there as a result of recombination.

We can estimate the voltage drop across the transport resistance. The charge current through a solar cell is limited by the absorbed photon current and has a maximum of 42 mA cm^{-2} for silicon in nonfocused solar radiation. For a doping concentration of $n_A = 10^{16}$ cm^{-3} and a mobility of $b_h = 470$ cm^2 $(Vs)^{-1}$, the conductivity is $\sigma_h = 0.75$ Ω^{-1} cm^{-1}. The voltage drop is then $1/e \,\text{grad}\, \eta_h = j_Q/\sigma_h = 56$ mV cm^{-1}, for a thickness of 300 μm, thus less than 2 mV. This is negligible compared with $(\eta_e + \eta_h)/e$, which is of the order of 1 V. We conclude that the transport resistance is negligible. The current through a pn-junction is, instead, limited by the reaction resistance.

We therefore have

$$\text{grad}\, \eta_h \approx 0 \quad \text{for} \quad x > -L_h$$

and

$$\text{grad}\, \eta_e \approx 0 \quad \text{for} \quad x < L_e$$

so that

$$\eta_e + \eta_h \neq f(x) \quad \text{for} \quad -L_h < x < L_e$$

It then follows that $\eta_e + \eta_h = eV$ where V is the voltage between the terminals for the n- and p-regions.

From the considerations above, it follows that the exponential function in the integrand in Equation 6.23 is constant over the range of the integration limits. This greatly simplifies the integration of Equation 6.23, and we obtain the current–voltage characteristic of the pn-junction as

$$j_Q = e\, G_h^0\, (L_e + L_h) \left[\exp\left(\frac{eV}{kT}\right) - 1\right] - e \int_{-L_h}^{L_e} \Delta G_h \, dx \tag{6.24}$$

An external short circuit ($V = 0$) defines the short-circuit current density j_{sc}

$$j_Q = -e \int_{-L_h}^{L_e} \Delta G_h \, dx = -e \int_{-L_h}^{L_e} \Delta G_e \, dx = j_{sc} \tag{6.25}$$

In the dark ($\Delta G_{e,h} = 0$) and for large negative voltages ($\exp(eV/kT) \ll 1$), we find the reverse saturation current j_S, which is independent of the voltage,

$$j_Q = -eG_{e,h}^0\, (L_e + L_h) = -j_S \tag{6.26}$$

The short-circuit current j_{sc} and the reverse saturation current j_S are the essential elements of the current–voltage characteristic of a pn-junction

$$j_Q = j_S \left[\exp\left(\frac{eV}{kT}\right) - 1\right] + j_{sc} \tag{6.27}$$

Using Equation 6.26, we can calculate the reverse saturation current for an ideal pn-junction, in which electron–hole pairs are produced only by the absorption of 300 K background radiation using the generation rate in Equation 3.54 and the lifetime from Section 3.6.3.

In real pn-junctions, we must also consider the electron–hole generation by nonradiative transitions as the reverse process of nonradiative recombination. The rate of generation, then, cannot be given in general form. However, it can be expressed in terms of the diffusion lengths, since the lifetimes that these contain represent a parameter that reflects the real recombination processes. For equilibrium between generation and recombination, the lifetime (of the minority carriers) together with the concentration of the minority carriers determines the generation rate, so that

$$G^0_{e,h} = R^0_{e,h} = \frac{n^p_e}{\tau_e} = \frac{n^n_h}{\tau_h}$$

From $L = \sqrt{D\tau}$ we obtain $\tau_e = L_e^2/D_e$ and $\tau_h = L_h^2/D_h$. Making the substitutions $n_e^p = n_i^2/n_A$ and $n_h^n = n_i^2/n_D$ then gives us the reverse saturation current

$$j_S = e n_i^2 \left(\frac{D_e}{n_A L_e} + \frac{D_h}{n_D L_h} \right) \tag{6.28}$$

This expression for the reverse saturation current is valid even for real pn-junctions in which recombination is predominantly nonradiative, when empirically determined values are used for the diffusion lengths of the electrons and holes. However, recombination at interfaces, e.g. with the contacts, which is a problem in real solar cells, is not contained in this description.

For the short-circuit current, only the photons absorbed within the diffusion lengths are of interest. The pn-junction must therefore be no more than a distance L_h away from the surface. In fact, the n-layer at the surface is chosen to be very thin, and its absorption can be neglected. It must also be taken into account that the probability for the absorption of photons varies with the photon energy $\hbar\omega$. The generation rate resulting from light incident onto the n-layer at $x = 0$ is found by integration over the incident photon current spectrum $dj_\gamma(\hbar\omega, x = 0)$. This yields

$$\Delta G_{e,h}(x) = \int_0^\infty \alpha(\hbar\omega)(1 - r(\hbar\omega)) \exp(-\alpha(\hbar\omega)x)\, dj_\gamma(\hbar\omega, 0)$$

The contribution to the short-circuit current in the interval $d\hbar\omega$ is

$$\begin{aligned} dj_{sc}(\hbar\omega) &= -e\,[1 - r(\hbar\omega)]\, dj_\gamma(\hbar\omega, 0)\alpha(\hbar\omega) \int_0^{L_e} e^{-\alpha x}\, dx \\ &= -e\,[1 - r(\hbar\omega)]\{1 - \exp[-\alpha(\hbar\omega)L_e]\}\, dj_\gamma(\hbar\omega, 0) \end{aligned} \tag{6.29}$$

$[1 - r(\hbar\omega)]\{1 - \exp[-\alpha(\hbar\omega)L_e]\}$ is the familiar absorptance $a(\hbar\omega, L_e)$ of a layer of thickness L_e for photons with an energy $\hbar\omega$. If the diffusion length is greater than the thickness of the solar cell, the absorptance for the actual thickness must be used.[1] The short-circuit current is given by the photon current absorbed within

1) In this case, a perfect hole membrane at the rear side of the cell must prevent the electrons from reaching the back contact where they would otherwise recombine, without contributing to the short-circuit current.

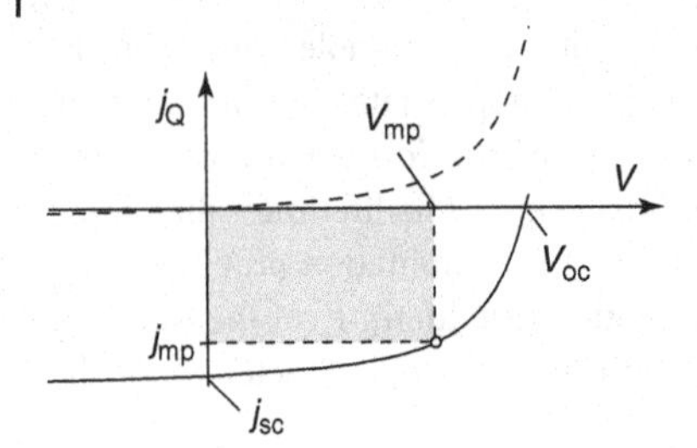

Figure 6.10 Charge current of the pn-junction in the dark (dashed line) and with illumination (solid line) as a function of the voltage. The sign of the voltage corresponds to the polarity of the p-region. The shaded rectangle represents the maximum power delivered by the illuminated pn-junction.

the shorter of the two lengths, the diffusion length of the electrons, or the thickness of the solar cell. For $L_e < d$, we get

$$j_{sc} = -e \int_0^{\infty} a(\hbar\omega, L_e)\, dj_{\gamma}(\hbar\omega, 0)$$

Figure 6.10 shows the current–voltage characteristic for the pn-junction in the dark and with illumination. The similarity with the characteristic of the chemical cell in Figure 6.4 is of no surprise since, for both, the current is limited by chemical reactions. The similarity is even greater. We leave it to the reader to show that the same linear current characteristic as shown in Figure 6.3 for the chemical cell is found for the semiconductor cell as well, if the current is plotted as a function of the product of the concentrations $n_e n_h$.

In addition to the short-circuit current, the open-circuit voltage V_{oc} is important. From

$$j_Q = j_S \left[\exp\left(\frac{eV_{oc}}{kT}\right) - 1\right] + j_{sc} = 0 \qquad (6.30)$$

we obtain

$$V_{oc} = \frac{kT}{e} \ln\left(1 - \frac{j_{sc}}{j_S}\right) \qquad (6.31)$$

It is essential for a large voltage that the reverse saturation current j_S is as small as possible. The generation rate in the dark $G^0_{e,h}$ has its lowest possible value, when electron–hole pairs are generated only by the absorption of 300 K radiation from the environment and therefore recombine only radiatively. In this case, the recombination rate is proportional to $\exp(eV/kT)$ resulting in the familiar current–voltage characteristic of the pn-junction. From the discussion of different recombination processes in Chapter 3, we know that the recombination rate has a different dependence on the difference of the Fermi energies for Auger recombination or for recombination via impurities, which leads to a different dependence of the current on the voltage.

6.5
pn-Junction with Impurity Recombination, Two-diode Model

The current–voltage characteristic calculated for the pn-junction in the previous section is based on the assumption that only radiative recombination takes place. This is the ideal situation that enables us to determine upper limits for the open-circuit voltage and the efficiency of a solar cell. In real solar cells, however, recombination via impurities predominates, as discussed in detail in Section 3.6.2. From there, we already know that impurities with an electron energy in the middle of the forbidden gap at $\varepsilon_{imp} = \varepsilon_i$ contribute particularly strongly to recombination. To simplify the evaluation of the influence of impurity recombination on the current–voltage relationship, we assume that electrons and holes have equal thermal velocities v and are captured with the same cross sections σ. With this simplification, Equation 3.82 leads to the relationship

$$R_{imp} = n_{imp}\sigma v n_i \frac{\exp\left[(\varepsilon_{FC} - \varepsilon_{FV})/kT\right] - 1}{\exp\left[(\varepsilon_{FC} - \varepsilon_i)/kT\right] + \exp\left[(\varepsilon_i - \varepsilon_{FV})/kT\right] + 2} \tag{6.32}$$

For a given voltage V and thus a given difference $\varepsilon_{FC} - \varepsilon_{FV} = eV$ between the Fermi energies, Figure 6.11 shows the recombination rate R_{imp} as a function of the mean value of the Fermi energies relative to the position ε_i of the impurity level assumed to be in the middle of the forbidden gap. It can be seen that the recombination rate has a pronounced maximum, when the Fermi energies lie symmetrically about the impurity level.

This behavior is understandable from Figure 6.12, which shows the potential distribution in a pn-junction at an applied voltage V and the position of the impurity levels relative to the Fermi energies. In the p-region, on the right, the impurity states are largely unoccupied, because their electron energies are above

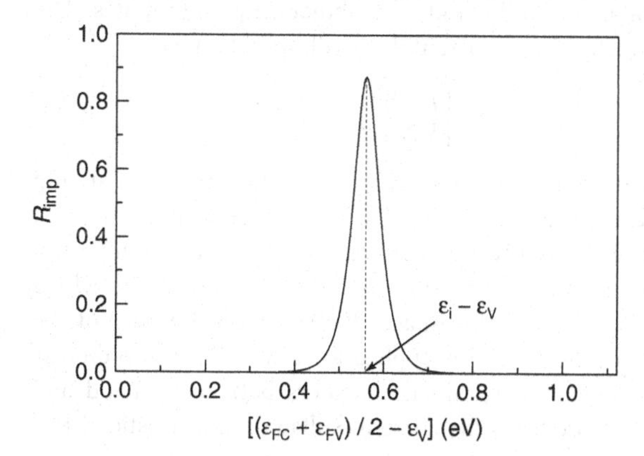

Figure 6.11 Impurity recombination rate R_{imp} as a function of the mean value of the Fermi energies $(\varepsilon_{FC} + \varepsilon_{FV})/2$ relative to the edge of the valence band ε_V in a semiconductor with an energy gap of $\varepsilon_G = 1.12$ eV calculated for an applied voltage of $V = 0.4$ V.

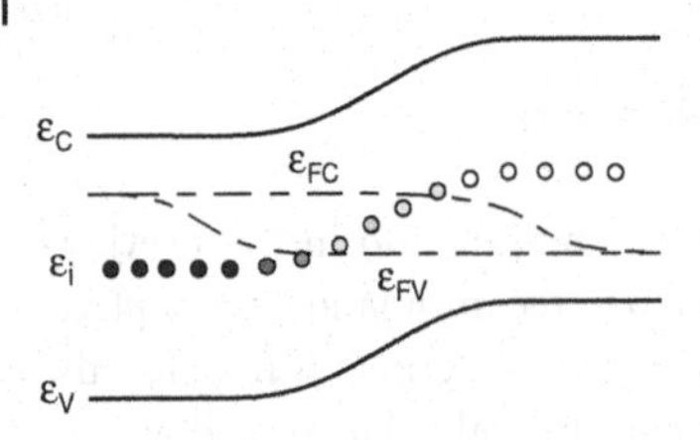

Figure 6.12 Potential distribution in a pn-junction with impurities at ε_i in the middle of the forbidden zone.

the Fermi energies, and in the n-region, on the left, they are largely occupied. In both regions, the recombination rate is small. The recombination rate is large only where the impurity level is in between the Fermi energies, that is, right at the pn-junction. There $\varepsilon_{FC} - \varepsilon_i = \varepsilon_i - \varepsilon_{FV} = eV/2$. This means that the recombination rate in Equation 6.32 is

$$R_{imp} = n_{imp}\sigma v n_i \frac{\exp(eV/kT) - 1}{2[\exp(eV/2kT) + 1]} \tag{6.33}$$

Let us further assume that this recombination rate is constant over the thickness w of the space charge layers and also consider that $\exp(eV/kT) - 1 = [\exp(eV/2kT) + 1]\,[\exp(eV/2kT) - 1]$. As a result, the impurity recombination leads to an additional charge current of

$$j_{Q,\,imp} = \frac{e w \sigma v n_{imp} n_i}{2}\left\{\exp\left(\frac{eV}{2kT}\right) - 1\right\} = j_{S2}\left\{\exp\left(\frac{eV}{2kT}\right) - 1\right\} \tag{6.34}$$

This current flows in addition to the current produced by band–band recombination, which determines the current over the region of the diffusion lengths. Because of the different voltage dependence, at small voltages, impurity recombination predominates and, at large voltages, band–band recombination predominates. The total current, including the short-circuit current due to illumination, is

$$j_Q = j_{S1}\exp\left\{\left(\frac{eV}{kT}\right) - 1\right\} + j_{S2}\exp\left\{\left(\frac{eV}{2kT}\right) - 1\right\} + j_{sc} \tag{6.35}$$

Except for the short-circuit current, we can visualize this total current as arising through two types of diodes connected in parallel. One diode with the reverse current j_{S1} in which only band–band recombination takes place, is connected in parallel with two diodes in series both having a reverse current j_{S2} and in which only impurity recombination occurs in the space charge region. Because of the series connection, only half of the voltage is applied to each of these two diodes. This summarizes the two-diode model, which considers both band–band and impurity recombination and reproduces the current–voltage characteristic of real pn-junctions relatively well.

A remark may be appropriate at this point. The exponential dependence of the current on the voltage in a pn-junction is often seen as being caused by the potential barrier between the n- and p-sides, which the carriers in the forward direction of the current have to surmount. In this description, it is very hard to understand that the carriers "see" only half of this barrier when they subsequently disappear by impurity recombination. We know, however, that the rise of the forward current is, instead, caused by the dependence of the recombination rate on the difference of the Fermi energies and it is no surprise that these dependencies are different for different recombination processes.

We have seen that the properties of even a homojunction, a pn-junction consisting of the same base material, can be interpreted in terms of the membrane model. The contributions to the current originate from chemical reactions between electrons and holes on the one hand, and photons and phonons on the other hand, occurring within a reaction volume extending from a diffusion length left of the junction to a diffusion length right of the junction. The materials outside the reaction volume, n-type on one side, p-type on the other side, serve as membranes, in which charge transport involves only the majority carriers. pn-junctions are not ideal for solar cells for the following reason. For the best performance of a solar cell, the recombination probability should be made as small as possible and the semiconductor material as pure as possible, allowing only for radiative recombination. We have learned in Section 3.6.3 that the radiative lifetime of electrons is 30 ms in p-type silicon doped with 10^{16} cm^{-3} acceptors, typical for crystalline silicon solar cells. For this lifetime, the diffusion length is $L_e \approx 1$ cm. In order to preserve the function of the hole membrane on the p-side that keeps the electrons away from the hole contact at the rear, the p-region of the solar cell would have to be thicker than that. The solar cell would then have to be much thicker than required to absorb the absorbable photons with $\hbar\omega \geq \varepsilon_G$, a waste of precious material. In Section 7.3, we see how this problem is solved. In any case, the membrane function is necessary to prevent minority carriers from reaching the wrong metal contact, although gradients of their electrochemical potentials exist to drive them into the wrong direction.

6.6 Heterojunctions

In Figure 5.6 we saw that in a homojunction there is transport in the wrong direction, that is, electrons flow to the contact on the p-side and holes to the n-side. The associated charge current by which the total current is reduced was neglected in the calculation of the current–voltage characteristic for the pn-junction. In fact, neglecting this contribution is justified only when measures are taken to eliminate this current or its cause, which is recombination at the interface with the metal contact. One possibility is shown in Figure 6.5. The electrons flow through an n-conductor out of the cell, and the holes through a p-conductor, both with a large energy gap. This results in very low concentrations of minority carriers in the membranes and so it is justified to neglect minority carrier currents in the wrong direction, namely, an electron current toward the p-side and a hole current toward the n-side. The structure of Figure 6.5 requires three different materials: an absorbing semiconductor in the middle, between two semiconductors with a larger energy gap and different electron affinities χ_e. Such combinations of different materials are known as *heterojunctions.*

In addition to preventing interface recombination at metal contacts, heterojunctions are important if it is not possible to prepare a pn-junction from a single material, a so-called homo junction. There are, in fact, many materials that can only be doped either n-type or p-type. This includes nearly all materials with an energy gap greater than 2.5 eV. This is one of the reasons why it is so difficult to manufacture blue-emitting light emitting diodes (LEDs), with the exception of GaN.

For solar cells and other electronic components, it is not sufficient to bring semiconductors with suitable energy gaps and electron affinities into contact. It is more important that the interfaces are as free as possible of states with energies in the forbidden gap, in order to eliminate additional recombination and also to prevent electrical charging due to preferred trapping of one carrier type. Material combinations fulfilling these conditions are very rare. One suitable combination is silicon–silicon dioxide, in which the silicon dioxide cannot be doped and therefore serves as an insulator. Nevertheless, we will see later that this combination is of great importance for silicon solar cells. Most other known combinations with a low density of interface states consist of III–V compounds and are based on gallium arsenide.

The distribution of the electrical potential and of the band edges can be determined for a heterojunction just as easily as for a normal pn-junction, if the interfaces are not charged, since this implies that the dielectric displacement $D = \epsilon\epsilon_0 E$ and the electrical potential φ are continuous over the interface. Because of the continuity of D, the field strengths E to the right and left of the interface differ by the ratio of the dielectric constants ε.

In Figure 6.13 we see two different semiconductors, semiconductor 1 on the left, which is n doped, and semiconductor 2 on the right, which is p doped. To determine the potential distribution in the dark, we begin in Figure 6.14 with semiconductor 1, for which – except for the possible occurrence of a space charge layer – the electrical potential is held constant to the left of the interface. In electrochemical equilibrium with the p-conductor, the Fermi energy has the same value everywhere. To the right of a space charge layer, possibly occurring at the interface, the band edges ε_C and ε_V and the electron affinities χ of semiconductor 2 are entered relative to the Fermi energies, with the same values as in Figure 6.13.

The distribution of the electrical potential is determined by the charge distribution and is independent of the electron affinity or the energy gap. Owing to the continuity of the electrical potential across the interface, in the case without charge in interface states, the resulting potential distribution is, apart from the different slope arising from the ratio of the dielectric constants, the same as for the homo-pn-junction in Section 6.4.

According to Equation 6.9, the electrical potential difference between the neutral regions of the n-conductor and the p-conductor is

$$e(\varphi^1 - \varphi^2) = \mu^1_{e,0} + kT \ln\left(\frac{n^1_e}{N^1_C}\right) - \mu^2_{e,0} - kT \ln\left(\frac{n^2_e}{N^2_C}\right) \tag{6.36}$$

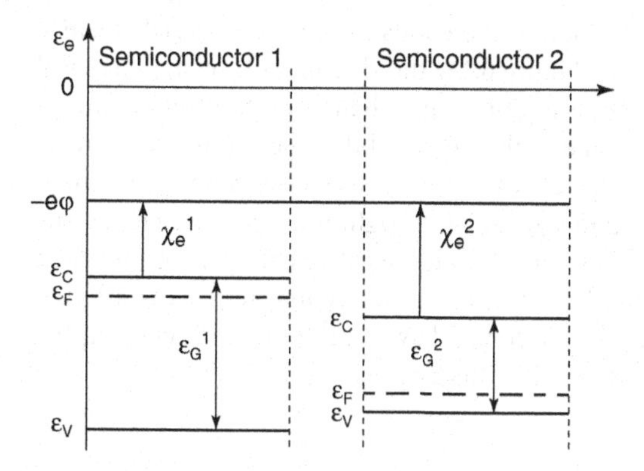

Figure 6.13 Two different semiconductors prior to making contact.

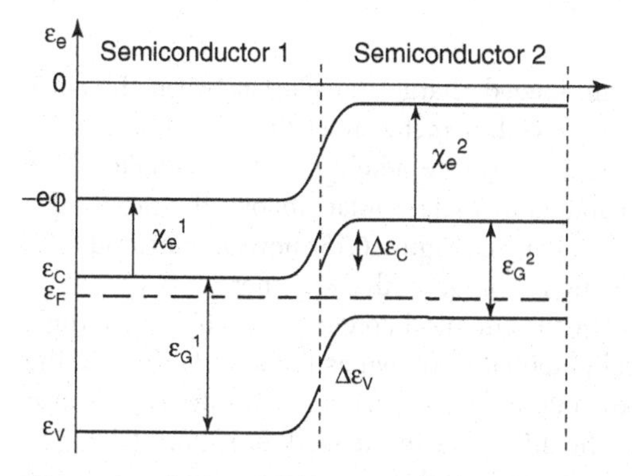

Figure 6.14 The two different semiconductors of Figure 6.13 in contact.

where the standard values of the chemical potential $\mu_{e,0}$ of the electrons are now given by the different electron affinities χ_e of the two semiconductors:

$$\mu_{e,0}^1 - \mu_{e,0}^2 = \chi_e^2 - \chi_e^1$$

The difference in the electron affinities also determines the discontinuity $\Delta\varepsilon_C$ in the conduction band edges and, together with the difference ε_G in the energy gaps, the discontinuity $\Delta\varepsilon_V$ of the valence band edges directly at the interface.

$$\Delta\varepsilon_C = \varepsilon_C^1 - \varepsilon_C^2 = \chi_e^2 - \chi_e^1$$
$$\Delta\varepsilon_V = \varepsilon_V^1 - \varepsilon_V^2 = \chi_e^2 - \chi_e^1 + \varepsilon_G^2 - \varepsilon_G^1$$

By selecting materials with suitable electron affinities, χ_e, and energy gaps, ε_G, as in Figure 6.5, it is possible to avoid discontinuities in the band edges for the majority carriers and produce them for the minority carriers, and, in this way, control the charge transport.

If interface states are present on one or both materials, the interface may be charged or may have a dipole moment or both. While outside the space charge regions from the interface, the position of the bands in the two materials relative to each other remain as shown in Figure 6.14, the distribution of the electrical potential φ within the space charge regions may change quite dramatically. A charge of one sign would add a spike to the potential distribution in Figure 6.14, upward for negative charge and downward for positive charge. A dipole moment would add a step, e.g. upward in going from left to right if the positive charge sits on the surface on the left of the interface and the negative charge is on the right surface. Since the position of the bands in the space charge regions is affected in the same way, charging of the interface will have a profound influence on the transport of electrons and holes across the interface.

6.7
Semiconductor–Metal Contact

Metal contacts should allow an unimpeded charge transport between the solar cell and an external load. The contact with a metal must therefore not cause a depletion of the majority charge carriers in the adjoining semiconductor. The potential distribution at the semiconductor–metal contact follows the same rules as at a semiconductor–semiconductor contact. Figure 6.15 shows a semiconductor and a metal before making contact and in contact with each other.

The metal is characterized solely by the chemical potential $\mu_{e,m}$ of its electrons. The absolute value of the chemical potential is known as the *work function* of the metal. In contact, an electrical potential difference arises, as for the case of two semiconductors, corresponding to the difference in the work functions. Owing to the high concentration of electrons in the metal, the charge distribution in the metal degenerates to a surface charge. The entire potential difference between the metal and the semiconductor, therefore, appears across the space charge layer of the semiconductor. In Figure 6.15, we have assumed that there is no charge at the contact interface other than the surface charge of the metal, so that the electrical potential is continuous across the interface. From this figure, we can

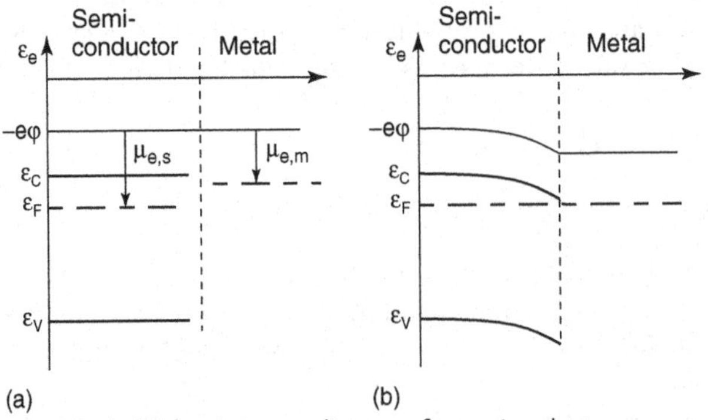

Figure 6.15 Schematic energy diagram of a semiconductor and a metal (a) before making contact and (b) in contact.

see that metals having a smaller work function than the semiconductor cause an accumulation of electrons in the adjacent semiconductor. Metals with a small work function, therefore, are favorable for the exchange of electrons and make good, so-called ohmic, contacts to n-type semiconductors. Conversely, metals with a large work function make good, ohmic contacts to p-type semiconductors. Electrons see a potential barrier from the Fermi energy of the metal to the conduction band of the semiconductor, equal to the difference between the work function of the metal and the electron affinity of the semiconductor. Heat is consumed when electrons cross the barrier in traveling from the metal to the semiconductor, resulting in a Peltier cooling effect. If, according to the direction of the charge current, electrons travel from the semiconductor to the metal, a Peltier heating effect is observed.

For contacts on covalent semiconductors, silicon in particular, the band bending in the semiconductor is found to depend less strongly on the work function of the metal than would be expected from the difference in the work functions. This is probably caused by the presence of surface states on the silicon surface, which are charged in contact with the metal. Together with the surface charge on the metal, the surface charge on the silicon forms a charge double layer, over which a potential step occurs, which depends on the dipole moment of this layer. The potential difference between the interior of the semiconductor and the interior of the metal is still given by the difference in the work functions, but it is now divided between the band bending of the space charge layer and the, usually unknown, potential difference across the dipole layer.

Contacts with good carrier exchange can also be prepared on the basis of another principle. If the semiconductor is very highly doped, at least in the vicinity of the contact, a majority carrier depletion layer caused by an inappropriate work function of a metal is only very thin. It can, in fact, for suitably high doping levels, be so thin that the majority charge carriers can tunnel through this potential barrier between the semiconductor and the metal. Aluminum has a smaller work function than p-doped silicon and is, therefore, not expected to be a good contact material for it. If, however, after deposition, the aluminum is allowed to diffuse at high temperatures into the silicon, where it forms acceptor states, a strongly p-doped layer results with a potential distribution as shown in Figure 6.16. The thin barrier in the valence band permits good hole exchange between the valence band of the silicon and the conduction band of the Al contact by tunneling. Generating a strongly doped p^+-layer in front of the metal contact also improves the membrane character of the p-region.

Table 6.1 Work functions and electron affinities of common semiconductors and metals.

	Si	GaAs	In	Ag	Al	Au	Pt
Work function (eV)			4.12	4.26	4.28	5.1	5.65
Electron affinity (eV)	4.01	4.07					

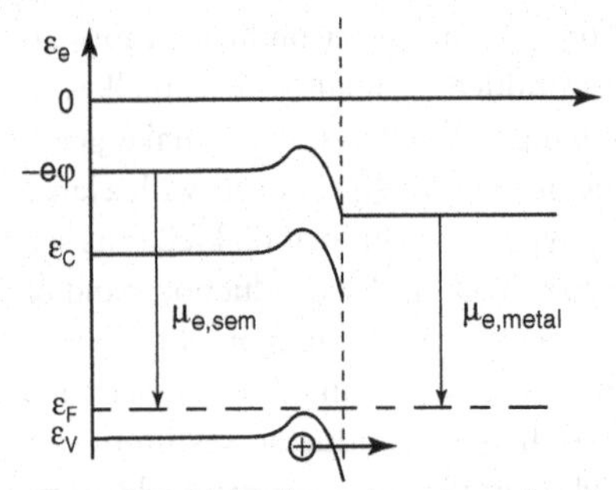

Figure 6.16 Holes from the valence band can tunnel through a thin potential barrier of a strongly p-doped depletion layer into the metal.

6.7.1 Schottky Contact

The basic principle of a solar cell, namely, having two membranes, one exchanging electrons and the other exchanging holes, can be fulfilled simply by a homogeneously doped semiconductor with two different metal contacts, one ohmic contact for the exchange of the majority carriers and a second that causes depletion of the majority carriers, and thus accumulation of minority carriers. This second contact is called a *Schottky contact*. For an n-conductor, its work function must be much greater than that of the semiconductor, and for a p-conductor much smaller.

Figure 6.17 shows the potential distribution at a Schottky contact on a p-conductor in the dark. It must, however, be noted that Schottky contacts are simple to prepare only for a few semiconductors. Furthermore, they have the disadvantage that the required unimpeded exchange of the minority carriers is unavoidably coupled with a high level of surface recombination at the metal contact. For solar cells, Schottky contacts play a role only in the testing of new materials. Past experience has shown that well-functioning contacts, assumed to be Schottky contacts, were in fact semiconductor heterojunctions formed by a chemical reaction of the metal

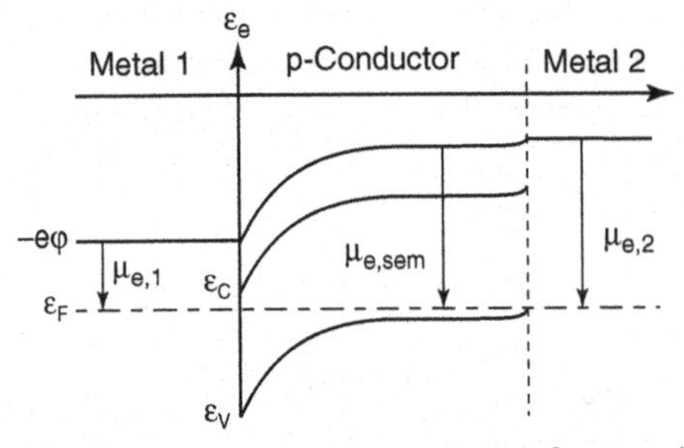

Figure 6.17 A metal with a small work function (left) forms a Schottky contact on a p-conductor with a larger work function. On the right side, a metal with a large work function forms an ohmic contact.

with the semiconductor. An example is the contact between copper (Cu) and n-type cadmium sulfide (CdS), where p-conducting Cu_2S is formed.

6.7.2
MIS Contact

To reduce the surface recombination at the Schottky contact, one can insert an oxide layer between the metal and the semiconductor. Even though it is an insulator, this is no serious obstacle when it is sufficiently thin to allow electrons or holes to tunnel through the thin potential barrier. However, the depletion of majority carriers in the semiconductor is then not as pronounced, because a part of the difference in the work functions is now occurring across the oxide layer and not in the semiconductor.

In metal insulator silicon (MIS) structures, [9] this disadvantage is compensated by exploiting a particular property of silicon dioxide. Impurity atoms bound to silicon dioxide (e.g. sodium) are frequently ionized, i.e. charged. Their charge is neutralized by the charge in the metal and in the semiconductor. If the oxide is positively charged at the interface to the semiconductor, holes are pushed away from the interface and electrons are accumulated. Figure 6.18 shows that a potential distribution similar to that of a Schottky contact results in the p-conductor, but now without the disadvantage of a high level of surface recombination.

6.8
The Role of the Electric Field in Solar Cells

The reader may find it confusing that the electric field that exists in the dark and, although somewhat reduced, also in the illuminated pn-junction, is of no significance for our understanding of the solar cell. The criterion for a solar cell structure is that electrons and holes are forced by membranes into different directions and that on their path their entropy is conserved. If this condition is

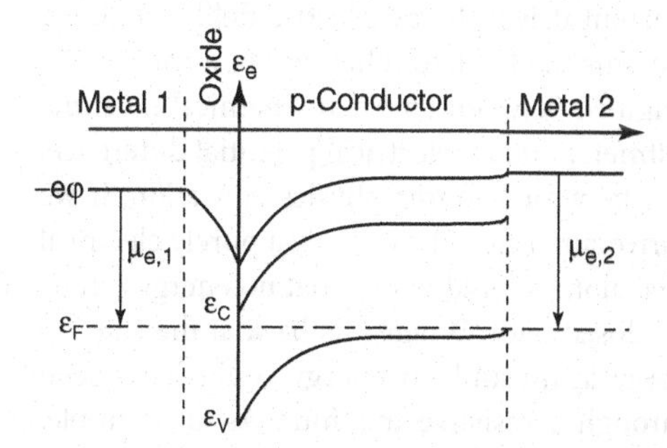

Figure 6.18 In a metal-insulator-silicon (MIS) structure, a very thin oxide layer between the metal and the silicon prevents surface recombination. The depletion of holes in the p-conductor is the result of a positive charge trapped in the oxide near the silicon.

fulfilled, in some structures, e.g. in a pn-junction of uniform material, an electric field will be present between the membranes. The direction of the short-circuit charge current in a pn solar cell agrees with the direction of this field. This seems to be sufficient to believe that it is also causing this current. To exaggerate somewhat, this is mere coincidence. It would be a completely unnecessary restriction to exclude structures for solar cells in which no electric field is present, but which have the membrane function incorporated and which fulfill the condition of conservation of entropy. The dye solar cell in Section 6.3 is a good example. The intimate mixture of electron membrane (TiO_2), dye, and hole membrane (electrolyte) on a nanometer scale, does not allow the formation of an extended space charge and of a field. Another example is given at the end of this section.

We frequently read that it is just the electric field of a pn-junction that supplies the driving force for the currents flowing during illumination. Let us take a closer look at this argument. If it were true that the charge carriers were driven by the field, then they would be continuously accelerated by the field and slowed down by collisions with the lattice. The field would perform work to keep them moving. There would have to be a source of energy present that continuously supplied energy to compensate for the energy dissipated during each collision of the charge carriers (see Section 5.1.3) in order to maintain a constant current. Such an energy source is, however, not present. We can see this best by comparison with a simple example.

Imagine a capacitor filled with a material that insulates in the dark and becomes conductive, as a result of the generation of electrons and holes, when illuminated. This capacitor is charged in the dark, and the voltage source is then disconnected. An electric field is present in the material between the plates of the capacitor. When this material is illuminated, a current will flow. Electrons flow to the positively charged plate and holes to the negatively charged plate. These particles gain energy from the field, which is subsequently dissipated by phonon scattering. However, the field is weakened by the charge transport, and the capacitor is discharged. The current vanishes after a dielectric relaxation time, and the electric field vanishes because the energy of the electric field stored in the capacitor during the charging process is used up. A steady state current driven by an electric field requires a continuous source of energy that maintains the field, in other words, a battery.

We have made some effort to distinguish between an electrochemical potential difference, which is measured by a voltmeter and an electrical potential difference, which is not. In fact, something must be wrong in our physics education, if we think that a DC current can at all be driven in a closed circuit by a purely electrical potential difference. The word potential alone should tell us that no energy can be gained by moving a charge along any closed path. When it is back at the starting point, it has the same electrical energy again and no energy could have been dissipated by a charge in moving through a resistive medium. As an example, Figure 6.19 shows water that can flow in a closed circuit. There is also a driving force, the gradient of the gravitational potential Φ, driving the water downwards. We know that the water in the pipe will not flow. If it would flow downhill on one side, it would have to flow uphill on the other side. But how does the water

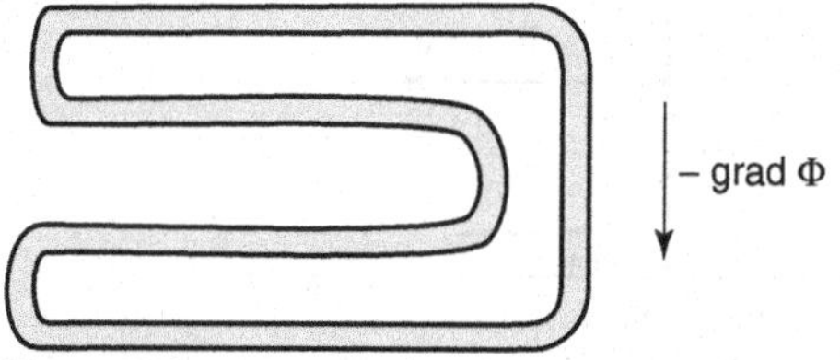

Figure 6.19 Water in a closed pipe does not flow, although the gradient of the gravitational potential Φ drives it downwards.

on the right "know" that it goes uphill on the left? A pressure gradient exists that compensates the gravitational force at every point. The same happens in a pn-junction. The gradient of the electrical potential is only maintained because it is compensated by the gradient of the chemical potential. Otherwise, the pn-junction would be discharged in the same way as the capacitor above, since there is no energy source to support the field.

A variation of the field as the driving force sees the driving force in the ability to dissipate energy. In Figure 6.20 we see that the energy of an electron in the conduction band of a pn-junction is larger on the p-side than on the n-side. By moving from the p-side to the n-side, the energy of the electron is reduced, that is, energy which came from the absorbed photon that created the electron–hole pair on the p-side. This energy reduction is largest under short-circuit conditions of an illuminated pn-junction and so, according to this argument, is the driving force. The error in this argument is that only free energy can be dissipated. We know that the free energy of an electron–hole pair is the difference of the two Fermi energies, which, in short circuit, is not large and is quite different from the energy change of an electron–hole pair when the electron is moved from the p- to the n-side and the hole from the n- to the p-side as is seen in Figure 6.21.

Energy, which is heat at the temperature of the solar cell cannot be dissipated, because the accompanying entropy is already at its maximum. The entropy per

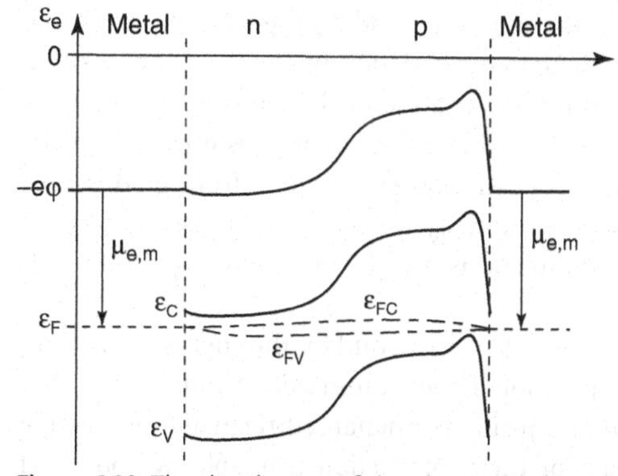

Figure 6.20 The distribution of the electrical potential φ in a pn-junction with metal contacts shows that a charge cannot gain energy from moving around a closed circuit.

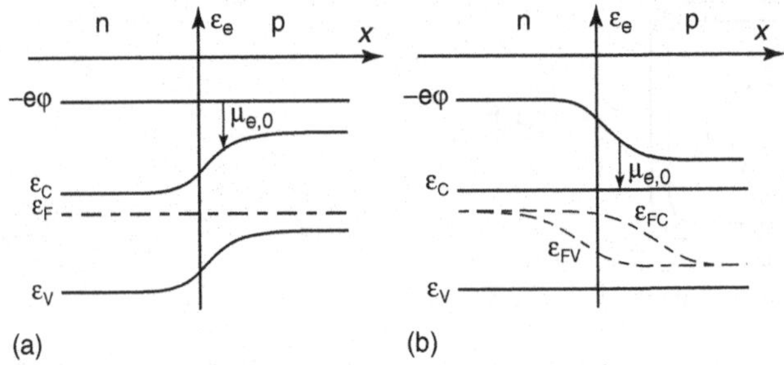

Figure 6.21 Potential distribution in a pn-junction, where the electron affinity $\chi_e(x) = -\mu_{e,0}(x)$ compensates the concentration-dependent part $kT \ln[n_e(x)/N_C]$ of the chemical potential. (a) In the dark and in electrochemical equilibrium.(b) Illuminated and at open-circuit.

electron can be read from Figure 3.16, which shows that, except for an amount of kT, the energy of the electron (hole) differs from its Fermi energy ε_{FC} ($-\varepsilon_{FV}$) by $\sigma_e T$ ($\sigma_h T$), which is heat at the temperature of the solar cell. We see that the reduction of the energy of an electron and a hole by separating them across a pn-junction is mainly a reduction of their heat along with a reduction of their entropy and has nothing to do with dissipation and a driving force. It is rather an exchange of heat and entropy with the phonon system. Under short-circuit conditions, the electrons and holes have gained their large entropy before moving across the pn-junction, because they were generated in an environment of low concentration. By driving the electrons and holes across the pn-junction by their proper driving forces (the gradients of their free energies), their entropy is given off to the phonon system and little entropy is generated when the gradients of the Fermi energies are small. This is the way by which entropy is given off to the environment, something that the solar cell has to do, in the same way as any other heat engine.

Figure 6.20 shows a pn-junction with contacts on the n- and p-sides. The contacts are made from the same metal as the wire (not shown) that connects them to ensure a short circuit. Outside the semiconductor, the electrical potential φ has the same value everywhere along the circuit, seen on the left and on the right in the figure. Inside the semiconductor it varies, owing to the variation in chemical composition (doping). A charge cannot benefit from this variation of the electrical potential by going around the circuit, since it goes uphill as much as it goes downhill in the electrical potential φ. Electrons and holes can, however, benefit from gradients of their electrochemical potentials, $\eta_e = \varepsilon_{FC}$ and $\eta_h = -\varepsilon_{FV}$, and the asymmetry of the membranes establishes preferred directions, for the electrons to go to the left and for the holes to go to the right.

To make perfectly clear that the electric field is irrelevant for the energy conversion in a solar cell, we construct a pn-junction in an admittedly cumbersome, but physically not forbidden, way where the p-side is uncharged relative to the n-side in the dark and no field is present. For the construction principle, we build on what we know about heterojunctions. According to Equation 6.36 the electrical potential difference between two bodies is determined by the difference in their work functions or by the difference in the chemical potentials μ_e of their electrons. Since the difference in the chemical potentials depends on the difference in the electron affinities and the ratio of the electron concentrations, it is conceivable to construct a pn-junction from continuously changing materials in such a way that all of them have the same energy gap and also have the same chemical potential for their electrons. In this structure, a spatial variation of the electron concentration, increasing from the p-side to the n-side, is compensated by a spatial variation of the electron affinity χ_e, also increasing from the n- to the p-region. In the dark, in electrochemical equilibrium, this structure has the potential distribution shown in Figure 6.21a.

When this structure is illuminated, an electric field is produced, which is oppositely oriented to the field in the normal pn-junction and increases with the illumination intensity. It would drive electrons to the p-region and holes to the n-region, if it were the only force, as seen on the right in Figure 6.21. Does this then mean that the current in this pn-junction flows in a different direction from that in a normal pn-junction?

The actual driving forces, namely, the gradients of the electrochemical potentials, are identical with those in a normal pn-junction. The pn-junction illustrated in Figure 6.21 therefore behaves exactly as a normal pn-junction and has the same current–voltage characteristic, even though the electric field is opposing the charge current. We know that the exponential dependence of the current on the voltage is caused by the dependence of the recombination rate on the difference of the Fermi energies and that it is therefore the same for the hypothetical pn-junction in Figure 6.21 as for a normal pn-junction.

That charge currents are flowing against an electric field in an energy source is, by the way, nothing unusual. We find this in every battery. While electrons flow from the minus pole to the plus pole in the external circuit, the continuity of the charge current requires that ions within the battery move against the electrical force, i.e. from the plus pole to the minus pole, if they carry a negative charge or in the opposite direction, if they are positive. They are able to do so, because a gradient of their chemical potential overcompensates the gradient of the electrical potential. This results in a gradient of their electrochemical potential – their only driving force – pointing in the right direction.

6.9 Organic Solar Cells

We have already discussed the dye solar cell in Section 6.3, but we did not mention one very important difference between organic and inorganic semiconductors concerning the generation of mobile charge carriers. In principle, in both material classes, the absorption of a photon does not directly lead to a free electron and a free hole since they attract each other because of their opposite charge. This entity in which electron and hole are still bound to each other by Coulomb forces is called an *exciton*. It is mobile and electrically neutral.

6.9.1 Excitons

A model for the exciton is the hydrogen atom as it was for the binding of an electron to a donor in Equation 3.25

$$\varepsilon_{exc} = \frac{m_{red}\, e^4}{2(4\pi\epsilon\epsilon_0)^2\hbar^2} \tag{6.37}$$

where the reduced mass of the exciton $m_{red} = m_e^* m_h^*/(m_e^* + m_h^*)$ is introduced to account for the movement of electron and hole around their common center of mass. Equation 6.37 gives the binding energy of the electron (or hole) in the ground state of the exciton. As known from the hydrogen atom, the exciton has a series of excited states above the ground state and below the ionization continuum where electron and hole are no longer bound to each other and are free to move. The binding energies following from Equation 6.37 are much smaller than kT at room temperature for most common inorganic semiconductors because of their large dielectric permittivity $\epsilon > 10$ and of effective masses of electrons and holes which are smaller than their real masses. As a consequence, excitons exist in inorganic semiconductors only at low temperatures, and at room temperature, electrons and holes are free to move after being generated by the absorption of a photon. No distinction is made between the minimum energy of absorbable photons and the minimum energy of free electron–hole pairs. This minimum energy is defined as the band gap for inorganic semiconductors.

This is different in organic semiconductors. They often consist of molecules that are only weakly bound to each other by van der Waals forces. As a result, they have a smaller dielectric permittivity and larger effective masses of electrons and holes. This raises the binding energy of the exciton and reduces its radius to a value comparable with an interatomic distance. For this size of the exciton, using bulk values for the dielectric permittivity and for the effective masses is no longer justified and the binding energies are even larger than predicted by Equation 6.37. The exciton binding energy in organic semiconductors may be as large as some tenths of an eV and therefore much larger than kT at room temperature. Even if excited above the ionization continuum by a photon with sufficient energy, electron and hole may quickly lose some of their kinetic energy by scattering with lattice vibrations and may then be trapped in their mutual attraction and form an exciton. Most absorption processes in organic semiconductors end up in excitons and the generation of free electrons and holes is the exception. In any case, we have to keep in mind that absorbable photons with the lowest energy, defining the absorption band gap, will not generate free charge carriers. The minimum energy of a free electron–hole pair, defining the free-carrier band gap, is larger than the optical band gap by the binding energy of the exciton. The exciton is a two-particle state with an integer spin and as such does not follow Fermi statistics but Bose–Einstein statistics. Its energy it cannot be well represented on a scale for electron energies. As a compromise, the exciton energy is often given with respect to the valence band edge. By measuring the exciton energy upward from the valence band edge

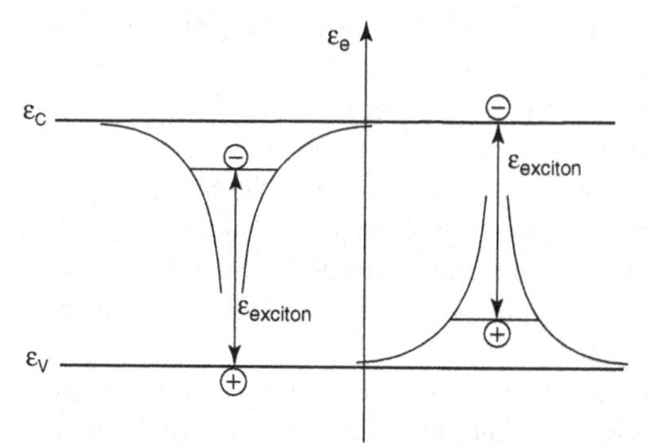

Figure 6.22 The exciton can be represented in two different ways. On the left, the electron is shown in the electric field of the hole and its energy (in the ground state of the exciton) is given relative to the energy of a free hole. The exciton binding energy is the energy difference from the ground state of the electron in the exciton to the conduction band edge ε_C. On the right everything is reversed as from the perspective of the hole in the field of the electron and its energy is relative to that of a free electron.

as shown on the left of Figure 6.22, it is represented as the energy of an electron bound to a free hole. Since in a material, the hole has as much reality as the electron, the exciton energy may as well be represented by the energy of a hole bound to a free electron. In this way, the exciton energy is counted downward from the conduction band edge as on the right in Figure 6.22. This representation is uncommon, but is very helpful in discussing the dissociation of excitons in organic materials.

The terminology for organic semiconductors is often different from the terminology used for inorganic semiconductors: the transfer of an electron to a molecule is described as a reduction of the molecule, and the transfer of a hole is an oxidation. Instead of the conduction band edge ε_C, we encounter the lowest unoccupied molecular orbital (LUMO), and instead of ε_V, we find the highest occupied molecular orbital (HOMO).

Advantages and Disadvantages

We start with the disadvantages of organic semiconductors. The rather weak van der Waals forces between the molecules of an organic semiconductor result in little overlap of electron wavefunctions from neighboring molecules. This causes narrow bands with large densities of states, large effective masses and, consequently, small mobilities of electrons and holes. The transport properties of organic semiconductors for electrons, holes and excitons are therefore rather poor.

On the other hand, organic semiconductors have advantages over inorganic semiconductors. The high densities of states in the bands result in large values of the absorption coefficient α on the order of 10^5 cm^{-1} for common direct transitions. To absorb most of the absorbable photons, layers of organic semiconductors do not have to be thicker than a fraction of 1 μm.

Another favorable property follows from the weak van der Waals forces between the molecules. States in the middle of the band gap that facilitate nonradiative recombination in the bulk of the semiconductor or on surfaces or interfaces have much lower densities than in inorganic semiconductors. Organic semiconductors, therefore, often have high fluorescence yields, which means that radiative recombination is more probable than nonradiative recombination. Organic semiconductors are therefore more efficient in transforming the energy from the Sun (heat) into chemical energy (of excitons) than inorganic semiconductors. And last not least, organic semiconductors need only moderate temperatures for their preparation, they are flexible, and may be cheap. They are, however, susceptible to oxygen and other gases and need tight encapsulation.

The favorable properties, however, do not make up for the poor transport properties. The exciton diffusion length is typically on the order of only 10 nm, i.e. much smaller than the penetration depth $1/\alpha$ of the absorbable photons.

For good organic solar cells, we have to solve two problems:

1. produce free electrons and holes from excitons
2. find a structure that is compatible with small transport distances and, nevertheless, complete absorption.

Dissociation of Excitons

It has been known for some time that even strongly bound excitons can dissociate into a free hole and a free electron at the interface with an appropriate different material. How this may work is explained by Figure 6.23. The exciton is represented as an electron bound to a free hole in material 1 on the left. The Coulomb interaction between electron and hole causes an electrical potential well $-e\varphi$ for the electron.

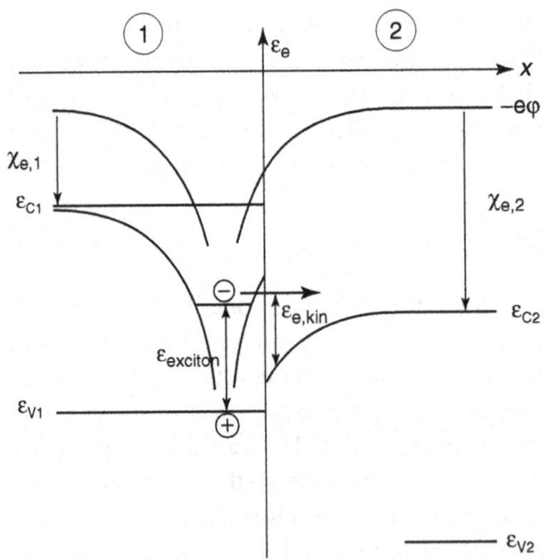

Figure 6.23 An exciton is shown in a material 1, the absorber, on the left as an electron bound to a free hole. The Coulomb potential well of the hole extends into material 2 on the right. This material has a larger electron affinity $\chi_{e,2}$ than material 1 on the left. As a result, the energy of a free electron $\varepsilon_e \geq \varepsilon_{C,2}$ is smaller in material 2 than the energy of the electron bound to a free hole in material 1, even if it is far away from the hole.

(The hole would see a totally different potential distribution, which is suppressed since the reference is the free hole.) The Coulomb well for the electron extends into material 2 on the right. Caused by a larger electron affinity $\chi_{e,2}$ of material 2, the energy of a free electron in this material is smaller than that of the bound electron in material 1. If the exciton in material 1 comes very close to the interface with material 2, the electron may tunnel (at constant energy) out of its bound state into a free state in the conduction band of material 2. Since the electron affinity χ_e is so much larger in material 2 than in material 1, the electron has enough kinetic energy $\varepsilon_{e,\mathrm{kin}}$ in material 2 to overcome the Coulomb attraction of the hole and become a really free electron far from the free hole remaining in material 1.

This process of exciton dissociation has a problem. If the electron loses too much of its kinetic energy $\varepsilon_{e,\mathrm{kin}}$ by scattering with acoustic vibrations while it is still within the Coulomb well, it becomes trapped at the interface and will stay there until some nonradiative recombination process occurs. If, however, the mean free path of electrons in material 2 is similar to the extension of the Coulomb well, the electron can leave the attraction of the hole without losing much energy. If that is possible, after the dissociation of the exciton, we have a free electron in material 2 and a free hole in material 1 with a combined energy that is nearly the same as the energy of the photon by which the exciton was generated in material 1. Under these favorable conditions, very little energy is lost in the dissociation of the exciton.

6.9.2 Structure of Organic Solar Cells

The combination of two different materials that allows an electron to tunnel from a bound state in the absorber into a free state in the second material also has the properties of a semipermeable membrane for electrons, if the second material does not offer states for free holes at their energy in the absorber as shown in Figure 6.23. In this combination, only the (bound) electrons may pass from the absorber into the electron collector, but the passage of the holes is blocked. The configuration in Figure 6.23 has the effect of the two different membranes that are necessary for solar cells with inorganic absorbers where free electrons and holes are directly generated by absorbed photons. After the dissociation of the exciton in Figure 6.23, the hole and the electron seem to be in media where there are no free recombination partners, no free electrons in the absorber, and no free holes in the electron collector. For good transport properties of a solar cell, the absorber should be p doped for good hole conductivity and the electron collector should have good electron conductivity. While the electron collector can then be directly connected to a metallic contact, the absorber should not be in contact with a metal because that would result in nonradiative recombination of the excitons when they come close to the metal.

In a more schematic drawing, Figure 6.24 shows an ideal configuration of an organic solar cell employing an additional hole membrane on the left of the absorber, which allows the holes to pass toward the left to the metal contact, which

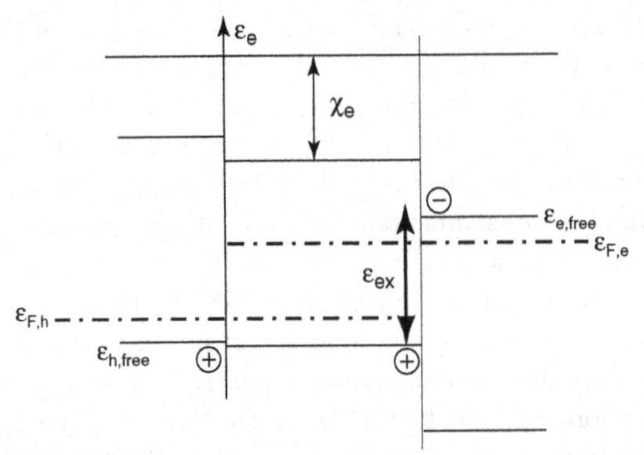

Figure 6.24 An exciton in the absorber in the middle dissociates into a free electron at the interface with the electron membrane on the right, which is then free to move to the metal contact on the right, represented by its Fermi energy $\varepsilon_{F,e}$. The free hole that remains in the valence band of the absorber after exciton dissociation can only leave the absorber through a hole membrane on the left and further on to a metal contact represented by $\varepsilon_{F,h}$.

is represented by its Fermi energy. This additional layer keeps the excitons away from the metal contact. In this ideal configuration, the sum of the energies of the free electron in the electron membrane on the right and the free hole in the hole membrane on the left in Figure 6.24 are almost equal to the energy of the exciton and of the absorbed photon by which it was generated. The use of organic materials in a solar cell does not mean that there is a principal energy loss associated with the dissociation of the excitons.

It would be consistent with the present discussion, if the electron collector on the right of Figure 6.24 would have the same bandgap as the absorber in the middle and therefore would absorb as well. Excitons in the electron collector could dissociate into a free electron in the electron collector and a free hole in the absorber in the middle. The energetic requirements for tunneling are just the same. The advantage would be that the material within an exciton diffusion length on both sides of the dissociating interface would contribute to the current, which could be twice as large as for absorption in the absorber only.

Although we have emphasized that illumination of an organic absorber generates excitons and no free carriers, Fermi energies for electrons ($\varepsilon_{F,e}$) and holes ($\varepsilon_{F,h}$) are shown for the absorber in Figure 6.24. It is true that the exciton has a large binding energy. Nevertheless, in steady state, there is chemical equilibrium between the excitons on one side and the electrons and holes on the other side in the absorber and this defines the values of the electron and hole Fermi energies. As always, one type of charge carrier can be fixed by doping, the concentration of the other type will adjust to make the sum of the chemical potentials of electrons and holes in the absorber (the difference of electron and hole Fermi energies) equal to the chemical potential of the excitons. The chemical potential of the excitons or of the electron–hole pairs determines the intensity of photon emission and the rate of radiative recombination by Equation 3.99 in exactly the same way as for inorganic semiconductors. Since in the limit of purely radiative recombination, the thermodynamic limits of the efficiency are approached, these limits are as valid for organic solar cells as for inorganic solar cells.

What we have discussed so far for a p-type absorber can be extended to an n-type absorber as well. In this case, exciton dissociation has to occur into a free hole in the valence band of a hole membrane with a free electron remaining in the conduction band of the absorber. This would be the right choice if the absorber exhibits good n-type conductivity, and if a hole membrane can be employed in which the free hole energy is smaller by the exciton binding energy than the free hole energy in the absorber.

Would it be a good idea to use the two concepts of exciton dissociation into a free electron in an electron membrane and into a free hole in a hole membrane simultaneously? The advantage would be that the absorber could be thicker. The excitons generated close to the front of the absorber would dissociate at one interface, while the excitons generated closer to the rear would dissociate at the other interface. Figure 6.25 shows such a configuration. Different from the previous concepts, both, free electrons and free holes exist in the absorber where they can recombine. Their concentrations, however, comply with the chemical equilibrium between electrons, holes, and excitons discussed earlier and the recombination rate is no larger than in the other concepts. In fact, the recombination rate is even smaller resulting from an inherent energy loss mechanism. The sum of the energies of free electrons in the electron membrane on the right and of free holes in the hole membrane on the left is now smaller than the exciton energy by the binding energy of the exciton. As a consequence, the difference of the electron and hole Fermi energies in the absorber, defining the voltage of the solar cell, is smaller than with only one exciton dissociating interface. Using two interfaces for exciton dissociation results in a principal loss of energy of at least the exciton binding energy and, as a consequence, in a loss of voltage in the solar cell.

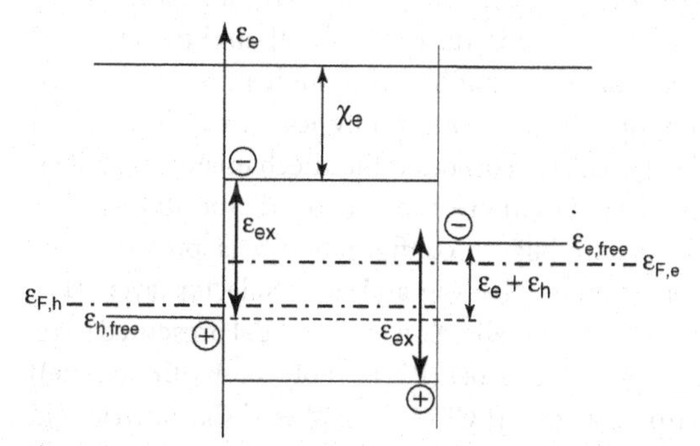

Figure 6.25 An exciton in the absorber in the middle may dissociate into a free electron at the interface with the electron membrane on the right or into a free hole at the interface with the hole membrane on the left. The absorber must support transport of the remaining holes from right to left and electrons from left to right.

(a)

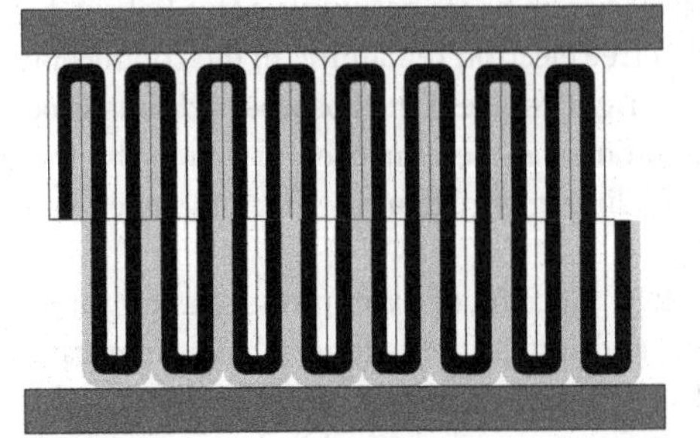

(b)

Figure 6.26 (a) In a planar configuration an organic absorber (black) is placed between a hole-conducting membrane (gray) and an electron-conducting membrane (white). Its thickness is less than the exciton diffusion length and therefore insufficient to absorb the absorbable light. (b) By folding the layer stack from (a), the absorption is increased without increasing the distance, the excitons have to diffuse. Contacts are added to the front and rear side of the folded layer stack. The front contact as well as the electron and hole collecting layers have to be transparent, of course.

The biggest problem of most organic materials is their short exciton diffusion length of only some nm. This is shorter by a factor of 10–100 than the penetration depth $1/\alpha$ of the light. A planar layer, in which all excitons have an interface for dissociation within reach, must be so thin that it can absorb only a small fraction of the incident light. Figure 6.26(a) shows such a thin absorber layer between a hole and an electron accepting collector. It also shows that folding these layers increases the optical thickness but leaves the condition for efficient exciton transport unchanged. Electrical contact is only made to the outside of the folded package provided the electron and hole collecting layers are sufficiently doped to warrant a small series resistance. With this geometry, a small exciton diffusion length and the required larger absorber thickness can be made compatible. Although more difficult to realize, efficient organic solar cells are not ruled out by physical principles.

Existing examples of organic solar cells incorporate the mechanisms described above. The first heterojunction solar cell where the exciton dissociation at the interface was incorporated in a planar bilayer configuration was presented by Tang [10]. More advanced versions contain electron and hole collector layers [11]. The bulk heterojunction solar cell is another realization of the model presented here. In an interpenetrating network, the p-type absorber (P3HT (Poly(3-hexylthiophene)) or similar) is mixed with an electron collector (PCBM ([6,6]-phenyl-C_{61}-butyric acid methyl ester) or other fullerenes) [12,13]. As a result, the electron collector may not be everywhere within reach of the excitons or may be in direct contact with the hole collector which may cause shunts. Whether the energetic alignment of all the energy levels involved conforms to the model above is still in question. There seems to be room for improvement.

6.10 Light Emitting Diodes (LED)

We have seen that light emission from solar cells is a result of the difference of the Fermi energies for electrons and holes. This difference may result from the generation of electron–hole pairs by illumination or from the injection of electrons and holes by applying a voltage without illumination. The first case belongs to the operation of a solar cell, the second to the operation of an LED. Both devices are based on the same structure, ideally a semiconductor placed between semipermeable membranes, for holes on one side and for electrons on the other side. If the recombination in a semiconductor is only radiative, the same device is an ideal solar cell and an ideal LED as well. If the thickness of the absorber/emitter is much larger than the penetration depth $1/\alpha$ of the photons and provided that the series resistance is negligible, operation as a solar cell and as an LED would be with maximum efficiency.

The question then arises, why is silicon, one of the best materials for solar cells, unsuitable for LEDs and on the other hand, why are organic LEDs quite efficient while organic solar cells are not?

For a solar cell, the incident intensity is fixed and we are interested in the charge current and voltage from the cell. For an LED, the voltage is fixed and we are interested in the charge current and in the emitted intensity. Two properties of nonideal devices are important: the absorptance and the contribution of nonradiative recombination.

We first discuss the effect of the absorptance, but assume that the recombination is purely radiative. For a solar cell, the absorbed photon current and with it the short-circuit current is proportional to the absorptance, but this absorptance $a(\hbar\omega)$ has to be counted only for a layer from which the minority carriers are able to reach the contacts without recombining. The thickness of this layer is equal to their diffusion length. For an LED, according to Equation 3.102, the emitted photon current is also proportional to the absorptance and again this absorptance is counted for a layer that is filled by minority carriers, when they are injected through their selective contact. The thickness of this layer is equal to the diffusion length. Although, the photon current, either absorbed or emitted, is directly proportional to $a(\hbar\omega)$ for both devices, an absorptance $a(\hbar\omega) < 1$ affects solar cells and LEDs differently. For a solar cell, nonabsorbed photons are lost and result in a proportionate loss of the short-circuit current and of the efficiency. For an LED, $a(\hbar\omega) < 1$ means that less photons are emitted than would be possible for the given voltage, but the charge current is accordingly smaller, since only the electron–hole pairs that recombine have to be replenished by the charge current. There is no direct loss involved. The photon current can be raised to the ideal value expected for $a(\hbar\omega) = 1$ by supplying a little more energy, by raising the voltage in Equation 3.102 by an amount of $kT/e \ln(1/a)$. So, if for an organic absorber/emitter, $a(\hbar\omega) = 0.01$, a solar cell would have an efficiency that is roughly smaller by a factor of 100 than the ideal efficiency and is unacceptably small, whereas for a LED, the small absorptance of this material is fully compensated by an additional voltage of only 120 mV.

The effect of nonradiative recombination is quite different. This time, we assume an ideal absorptance $a(\hbar\omega) = 1$. The short-circuit current of a solar cell is then ideal. Additional recombination by nonradiative transitions, however, reduces the steady-state carrier concentration under open-circuit conditions from its ideal value by a factor that is given by the total recombination rate divided by the radiative recombination rate. For fairly good silicon solar cells, for instance, recombination is 1000 times more probable in nonradiative transitions than in radiative transitions. The minority carrier concentration is reduced by a factor of 1000 and the open-circuit voltage is smaller than its ideal value by $kT/e \ln(1000)$, i.e. by about 180 mV. In contrast, the emitted intensity of this solar cell operated as an LED with a fixed applied voltage is unaffected by nonradiative recombination and is the same as for ideal properties. The charge current, however, which is needed to supply all the carriers that recombine is larger by a factor 1000 than for an ideal device. The efficiency of this silicon-LED is therefore reduced by the same factor 1000 and is unacceptably small.

In conclusion, we see that for solar cells, the efficiency is proportional to the absorptance, but depends only logarithmically on the nonradiative recombination probability. Small absorptivities are prohibitive, whereas some nonradiative recombination can be tolerated. For LEDs, the charge current is directly proportional to the total recombination rate and a large proportion of nonradiative recombination is prohibitive, but small absorptivities can be tolerated, since they are easily compensated by raising the applied voltage by a small amount.

6.11 Problems

6.1 What happens to the reverse saturation current density j_S of a pn-junction if the lifetimes of electrons and holes are each improved by a factor of 4.

6.2 Calculate the width of the space charge region (depletion zone) of a pn-junction in the one-dimensional Schottky model for an applied voltage of $V_a = -2.3$ V. Use $n_D = 10^{19}\ \text{cm}^{-3}$, $n_A = 10^{16}\ \text{cm}^{-3}$, and a dielectric function $\epsilon = 11.9$.

6.3 Calculate the capacitance of the pn-junction in the last problem without external voltage and for an applied reverse bias of 4 V. From the experimental determination of the capacitance as a function of the applied voltage it is possible to derive the diffusion voltage and, under certain conditions, also the doping concentration. How can that be done?

6.4 Assume a Si sample with acceptor-type impurity states with $\tau_{h,\min} = \tau_{e,\min} = 10^{-6}$ s. Calculate the separation of the quasi-Fermi energies $\varepsilon_{FC} - \varepsilon_{FV}$ for a temperature of $T = 300$ K as a function of the impurity level ($\varepsilon_V \leq \varepsilon_{imp} \leq \varepsilon_C$). Consider intrinsic Si and two different p-doping regimes with corresponding Fermi energies in the dark of $\varepsilon_F^0 = \varepsilon_i$, $\varepsilon_F^0 = \varepsilon_i - 0.2$ eV and $\varepsilon_F^0 = \varepsilon_i - 0.4$ eV for a generation rate

(a) $G = 10^{18}$ cm^{-3}s^{-1}

(b) $G = 5 \times 10^{20}$ cm^{-3}s^{-1}.

Use $N_C = 3 \times 10^{19}$ cm^{-3}, $N_V = 10^{19}$ cm^{-3} and $\varepsilon_G = 1.12$ eV. The electrons and holes shall be distributed homogeneously throughout the sample due to large diffusion lengths.

Plot a graph of the results as a function of the impurity level.

7
Limitations on Energy Conversion in Solar Cells

In connection with the derivation of the current–voltage characteristic, we have neglected a voltage drop over the transport resistances as being negligibly small. With this approximation, the voltage V at the contacts of a sufficiently doped solar cell is given by the separation of the Fermi energies, $eV = \varepsilon_{FC} - \varepsilon_{FV}$. In addition, we did not consider currents of minority carriers flowing in the wrong direction, in spite of the large gradients of their Fermi energies, because of the small conductivity of the minority carriers in regions that function as semipermeable membranes for the majority carriers.

With this approximation, all electrons and holes produced by the illumination, which do not recombine, contribute to the charge current. For homogeneous excitation at short circuit, the electron–hole pairs contributing to the charge current are those which are produced within the diffusion lengths. This approximation is justified especially in structures such as in Figure 6.5, in which the discontinuities at the band edges prevent the flow of minority charge carriers to the wrong side.

A solar cell structure consisting of an absorber with n- and p-type membranes supplies a charge current at a voltage V with $eV = \varepsilon_{FC} - \varepsilon_{FV} = \mu_e + \mu_h$. It supplies electrical energy equal to the chemical energy $\mu_e + \mu_h$ per electron–hole pair. This structure is able to convert the chemical energy produced by illumination of a semiconductor absorber *entirely* into electrical energy.

7.1
Maximum Efficiency of Solar Cells

The maximum energy current delivered by a solar cell is given by the largest rectangle fitting under the current–voltage characteristic, as shown in Figure 6.10. It defines the "maximum power point" for the charge current density j_{mp} and the voltage V_{mp}. For a given current–voltage characteristic, it is therefore important to have an algorithm to find the maximum power point.

Independent of the form of the characteristic, the functional relationship between j_Q and V, the condition for maximum power yields

$$d(j_Q V) = dj_Q V + j_Q dV = 0$$

Physics of Solar Cells: From Basic Principles to Advanced Concepts. Peter Würfel

ISBN: 978-3-527-40857-3

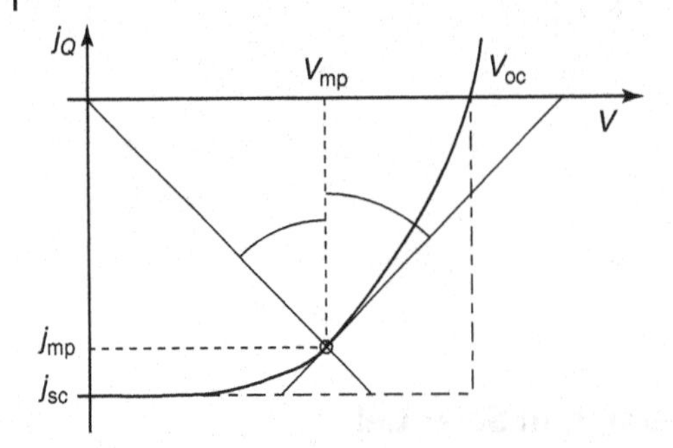

Figure 7.1 Geometrical construction of the maximum power point.

and thus

$$\left(\frac{dj_Q}{dV}\right)_{mp} = -\left(\frac{j_Q}{V}\right)_{mp} \tag{7.1}$$

This relationship is illustrated geometrically in Figure 7.1. From this construction, the maximum power point occurs where the tangent to the characteristic makes the same angle with a vertical line as the line connecting the coordinate origin and the maximum power point.

For a certain form of the characteristic, namely for radiative recombination, in Equation 6.27

$$\frac{dj_Q}{dV} = j_S \frac{e}{kT} \exp\left(\frac{eV_{mp}}{kT}\right) = -\frac{j_{mp}}{V_{mp}} \tag{7.2}$$

With

$$j_{mp} = j_S \left\{\exp\left(\frac{eV_{mp}}{kT}\right) - 1\right\} + j_{sc}$$

and

$$\frac{j_{sc}}{j_S} = 1 - \exp\left(\frac{eV_{oc}}{kT}\right)$$

we find from Equation 7.2

$$V_{mp} = \frac{kT}{e}\left\{\exp\left(\frac{e(V_{oc} - V_{mp})}{kT}\right) - 1\right\} \tag{7.3}$$

Solutions of this equation, and thus the maximum power point, can be found by numerical techniques. With the fill factor

$$FF = \frac{j_{mp} V_{mp}}{j_{sc} V_{oc}} \tag{7.4}$$

we define a measure for how well the maximum power rectangle fits under the characteristic. For only radiative recombination $j_{sc}\,V_{oc}$ represents the chemical energy current emitted by the photons in the open-circuit state, in which all electrons and holes must recombine.

An approximate value for the fill factor can be obtained from Equation 7.3. This gives us

$$V_{mp} = V_{oc} - \frac{kT}{e} \ln\left(1 + \frac{eV_{mp}}{kT}\right) \approx V_{oc} - \frac{kT}{e} \ln\left(1 + \frac{eV_{oc}}{kT}\right)$$

Since the logarithm depends only weakly on its argument, we have substituted V_{oc} for V_{mp} in the logarithm. With this result for V_{mp}, we derive the current j_{mp} at the maximum power point from the characteristic and find an approximate fill factor

$$FF = \frac{eV_{oc}/kT - \ln\left(1 + eV_{oc}/kT\right)}{1 + eV_{oc}/kT} \tag{7.5}$$

Values of the fill factor of good solar cells are between 0.8 and 0.9.

Maximum Short-circuit Current

A large short-circuit current requires a solar cell as thick as possible to maximize its absorptance. With antireflection coatings, we can theoretically reduce the reflectance to $r = 0$. For a cell with large thickness and at the same time a large diffusion length, the absorptance over the diffusion length is given by $a(\hbar\omega \geq \varepsilon_G) \approx 1$. The short-circuit current produced by the absorbed photon current is then

$$j_{sc} = -e \int_0^\infty a(\hbar\omega)\, dj_{\gamma,\,Sun}(\hbar\omega) = -e \int_{\varepsilon_G}^\infty dj_{\gamma,\,Sun}(\hbar\omega) \tag{7.6}$$

Maximum Open-circuit Voltage

The open-circuit voltage V_{oc} defines the separation $\varepsilon_{FC} - \varepsilon_{FV}$ of the Fermi energies at which recombination is in equilibrium with electron–hole generation throughout the entire cell. Owing to the exponential decay of the photon current density within the semiconductor, the rate of generation (per volume) is greatest at the surface. The electrons and holes produced are distributed by diffusion more or less uniformly over the thickness, depending on the diffusion length. The recombination rate is then equal to the averaged generation rate everywhere. When the thickness of the solar cell is reduced and surface recombination is prevented, the recombination rate (per volume) and with it $\varepsilon_{FC} - \varepsilon_{FV}$ must increase, because the averaged generation rate increases with decreasing cell thickness. The open-circuit voltage reaches a maximum when the thickness of the cell goes to zero. The open-circuit voltage increases only slightly, however, with decreasing thickness and this in no way compensates for the short-circuit current loss. In any case, a cell should not be made unnecessarily thick, and not merely to save on material. The optimum thickness is reached when a further increase in the thickness causes as much additional recombination at maximum power as it provides additional generation by more absorbed photons.

The greatest voltage is achieved if there is only radiative recombination, and then, surprisingly, the thickness of thick solar cells no longer plays a role. It is true that the total recombination rate increases in proportion to the volume. In thick cells, however, a large part of the photons produced by recombination does not reach the surface and is reabsorbed, producing electron–hole pairs again. The effective recombination rate integrated over the entire cell is equal to the photon current $j_{\gamma,\mathrm{emit}}$ emitted through the surface, which reaches a saturation value in thick cells when the absorptance reaches its maximum value $a = 1 - r$, which no longer depends on the thickness.

If the diffusion lengths are large compared with the thickness, the electrons and holes are uniformly distributed over the volume of the cell. For such a homogeneous distribution, we have in the approximation to the generalized Planck equation

$$j_{\gamma,\mathrm{emit}} = \int_0^{\infty} a(\hbar\omega)\, \mathrm{d}j_{\gamma}^{0}(\hbar\omega)\, \exp\left(\frac{\varepsilon_{FC} - \varepsilon_{FV}}{kT}\right) \qquad (7.7)$$

For the maximum short-circuit current, $a(\hbar\omega \geq \varepsilon_G) = 1$, and replacing $\varepsilon_{FC} - \varepsilon_{FV}$ by eV, the total rate of radiative recombination is

$$j_{\gamma,\mathrm{emit}} = \exp\left(\frac{eV}{kT}\right) \int_{\varepsilon_G}^{\infty} \mathrm{d}j_{\gamma}^{0}(\hbar\omega) \qquad (7.8)$$

independent of the thickness.

Defining the charge current from the p-region to the n-region (toward the left in Figure 6.5) as positive, the charge current delivered by the solar cell is

$$j_Q = \mathrm{e} j_{\gamma,\mathrm{emit}}(V) - \mathrm{e} j_{\gamma,\mathrm{abs}} \qquad (7.9)$$

or

$$j_Q = \mathrm{e}\left[\exp\left(\frac{eV}{kT}\right) - 1\right] \int_{\varepsilon_G}^{\infty} \mathrm{d}j_{\gamma}^{0}(\hbar\omega) - \mathrm{e}\int_{\varepsilon_G}^{\infty} \mathrm{d}j_{\gamma,\mathrm{Sun}}(\hbar\omega) \qquad (7.10)$$

Since the spectra of the 300 K background radiation $\mathrm{d}j_{\gamma}^{0}(\hbar\omega)$ and of the Sun $\mathrm{d}j_{\gamma,\mathrm{Sun}}(\hbar\omega)$ (outside the atmosphere *AM*0 and on the surface of the Earth *AM*1.5) are known, the maximum energy current $(j_Q V)_{\max} = j_{mp} V_{mp}$ can be determined from Equation 7.10 and, from this, the efficiency

$$\eta = \frac{j_{mp} V_{mp}}{\int_0^{\infty} \hbar\omega\, \mathrm{d}j_{\gamma,\mathrm{Sun}}(\hbar\omega)} \qquad (7.11)$$

7.2 Efficiency of Solar Cells as a Function of Their Energy Gap

The short-circuit current of a solar cell depends on the absorbed photon current. It is a maximum for a semiconductor with an energy gap $\varepsilon_G = 0$ and decreases with increasing ε_G. The open-circuit voltage V_{oc} is, however, zero for $\varepsilon_G = 0$ and increases with increasing energy gap. The efficiency η is therefore zero at $\varepsilon_G = 0$ and at $\varepsilon_G \to \infty$. Somewhere in between is its maximum. From Equations 7.10 and 7.11, we can calculate the efficiency η as a function of the energy gap ε_G if only radiative recombination takes place in the case of thick cells, in which $a(\hbar\omega < \varepsilon_G) = 0$ and $a(\hbar\omega \geq \varepsilon_G) = 1$.

Figure 7.2 gives the result for the *AM*0 spectrum outside the atmosphere, and Figure 7.3 for the *AM*1.5 spectrum on the surface of the Earth. We find a broad

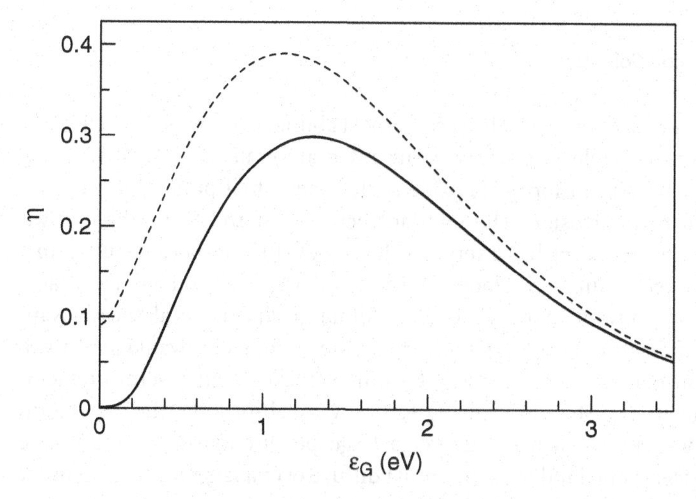

Figure 7.2 Efficiency of solar cells with radiative recombination only as a function of their energy gap for the *AM0* spectrum, non-concentrated (solid line) and for full concentration (dashed line).

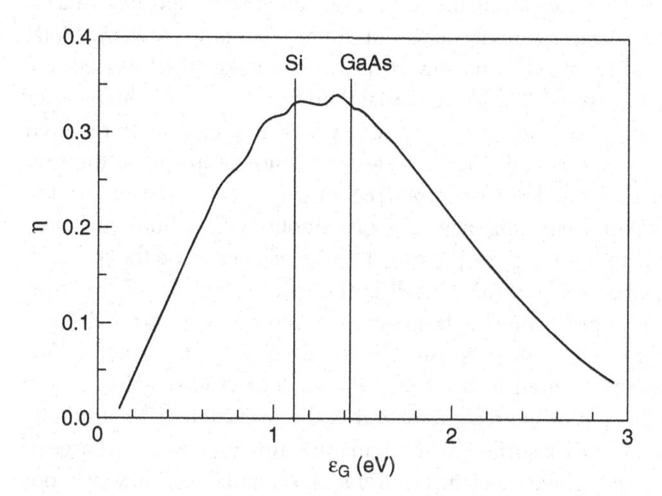

Figure 7.3 Efficiency of solar cells with radiative recombination only as a function of the energy gap for the *AM1.5* spectrum.

maximum, which indicates that semiconductors with an energy gap ε_G between 1 and 1.5 eV are suitable for solar cells. For the *AM1.5* spectrum, the maximum efficiencies are greater than for the *AM0* spectrum, because absorption in the atmosphere mainly eliminates photons with $\hbar\omega < 1$ eV, which cannot be utilized by solar cells with optimal band gaps. Silicon and gallium arsenide are especially well suited for the *AM1.5* spectrum.

7.3
The Optimal Silicon Solar Cell

Silicon has many advantages. It is the second most abundant element in the Earth's crust and is thus available in practically unlimited amounts. Silicon is not toxic. On exposure to air, silicon forms an oxide surface layer that protects it fully and prevents any further corrosion. The interface between Si and SiO_2 when grown under clean-room conditions has a very low density of surface states resulting in a very low surface recombination velocity. With $\varepsilon_G = 1.12$ eV, silicon has a favorable energy gap for the conversion of solar energy. Although silicon has all these advantages, with its indirect optical transitions it has the serious disadvantage of weak absorption. Consequently, silicon must be much thicker than a semiconductor with direct transitions. Moreover, owing to the weak absorption, the generation of electron–hole pairs is distributed over a large penetration depth $1/\alpha$ of the photons, and at least one kind of carrier must diffuse over a large distance in order to reach the contact. This implies that this carrier type must also have a large diffusion length and lifetime. Because of the poor absorption, not only is more silicon needed but it must also have a higher purity than if the optical transitions were direct.

In the usual structure of a solar cell, the contacts are applied to opposite surfaces. For the illuminated surface, a nontransparent contact poses a problem. Metal contacts are therefore arranged in narrow strips, in comblike structures leaving most of the surface uncovered. The charge must then flow in the membrane layer toward the contact strips parallel to the surface. A high doping level is required in this layer to avoid an intolerably large series resistance. Adjacent to the thin membrane layer is a weakly doped absorber region over the greatest part of the solar cell thickness. Since electrons have a greater mobility than holes and thus a greater diffusion length for a given lifetime, they are chosen to be the minority carriers. The large middle region of the cell is therefore p doped and the front surface is strongly n doped, with the designation n^+, to function as an electron membrane. To minimize the loss of electrons by surface recombination at the back contact, strong p-doping is used in front of the back contact to establish a hole membrane. The reduction of the recombination at the rear contact is commonly attributed to the so-called back surface field originating from the negative charge of the p^+-doped region, thought to repel the electrons. This repulsion is, however, not recognizable in the total force (grad η_e) and the smaller recombination probability is instead due to the reduced concentration of electrons in the p^+ layer.

Owing to the high conductivity of the rear p^+ layer a metal contact is not required over the entire rear surface. The areas of the front and rear surfaces without metal contact are covered with a passivating layer, silicon dioxide or silicon nitride, in order to reduce the rate of surface recombination. On the rear surface, the oxide layer is covered by a metal mirror layer that reflects the photons not yet absorbed and thus enhances absorption in the cell. The oxide or nitride layer on the front side is in the form of a $\lambda/4$ layer to reduce reflection for the wavelength range around λ in the infrared and red part of the spectrum, where the absorbable photon current

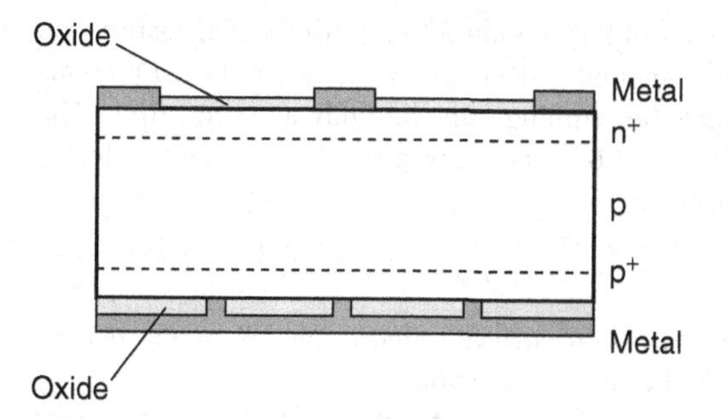

Figure 7.4 Cross-section of a silicon pn solar cell.

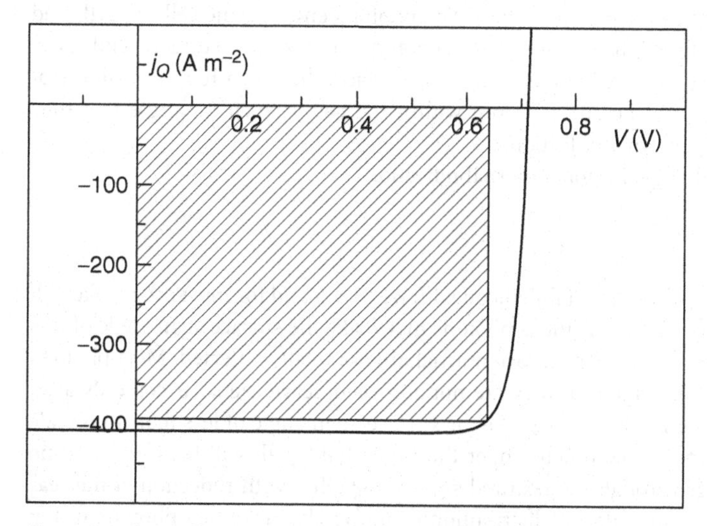

Figure 7.5 Charge current j_Q as a function of the voltage V for a Si solar cell with a thickness of 400 μm illuminated by the *AM*1.5 spectrum. The maximum power given by the rectangle corresponds to an efficiency of 25%.

has its maximum. Reflection in the blue is less reduced, which gives silicon solar cells their characteristic blue appearance.

Figure 7.4 illustrates a cross section through this structure. The current–voltage characteristic in Figure 7.5 was calculated for the *AM*1.5 spectrum, assuming that the front side is nonreflecting and only the unavoidable radiative recombination and Auger recombination corresponding to the necessary doping concentrations are considered, but not surface recombination. This cell has an efficiency of 25%.

7.3.1
Light Trapping

The absorptance of a body increases as the reflectance is reduced and the path of the photons within the body becomes longer. This triviality enables us to consider still another way to improve the absorptance than by an antireflection coating and a large thickness. The reflectance of a body decreases when the reflected photons are deflected in such a way that they impinge on the body a second time. The pyramid-shaped structure of Figure 7.6 makes this possible. For light reflected twice, the total reflectance is given by

$$r_{\text{total}} = r_{\text{in}}^2$$

A surface with 10% reflection as a planar surface reflects only 1% in a structure, where each reflected photon hits the surface a second time.

In addition, as a result of the textured surface, together with a reflecting rear surface, the light path in a solar cell is considerably increased compared with normal incidence on a planar surface. The photons entering the cell are deflected in an oblique direction because of refraction at the textured surface. And, even more important, after reflection at the rear surface, there is a high probability of their impinging on the surface from within the cell at such an angle that they experience total internal reflection.

At the critical angle for total internal reflection

$$\sin \alpha_{\text{T}} = \frac{1}{n_{\text{Si}}}$$

Since the index of refraction for silicon with $n_{\text{Si}} = 3.5$ and for most other solar cell materials is very large, only those photons that strike the surface at an angle of less than $\alpha_{\text{T}} = 16.6^\circ$ to the local surface normal are not totally reflected. Most photons are therefore trapped and, if they are not absorbed, leave the solar cell only after multiple reflections when they strike the surface at an angle of less than 16.6°. We can easily estimate how much longer the mean light path will be, if we assume that the passage through the textured surface, together with reflection at the rear surface, leads to an isotropic distribution of the weakly absorbed photons in the solar cell. The photons that are emitted through the surface then have the same angular distribution with which a black-body Lambertian surface emits photons. For this angular distribution, the photons leaving through the front surface cover an effective solid angle of π outside the solar cell (for the assumed large index of refraction), as described in Section 2.1.4. The isotropic distribution internally

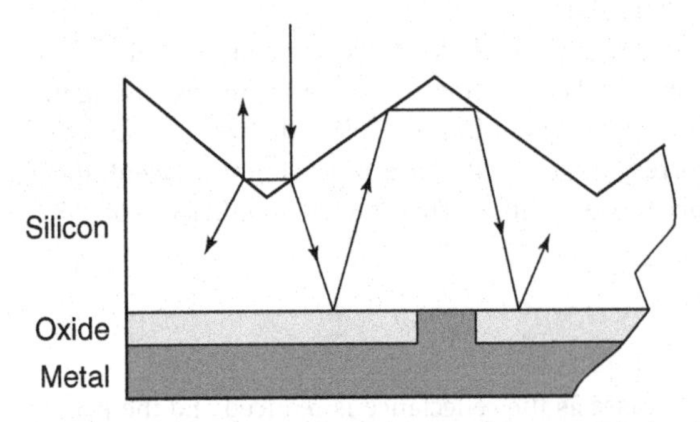

Figure 7.6 Surface texture for reducing reflection and increasing the length of the light path.

fills the solid angle 4π, in which the photon current density per solid angle $j_{\gamma,\Omega}$ is greater by a factor of n^2 than outside the solar cell.

For a Si cell of thickness L with a surface area A the absorptance is defined as

$$a(\hbar\omega) = \frac{I_{E,\mathrm{abs}}(\hbar\omega)}{I_{E,\mathrm{inc}}(\hbar\omega)} = \frac{I_{\gamma,\mathrm{abs}}(\hbar\omega)}{I_{\gamma,\mathrm{inc}}(\hbar\omega)}$$

The absorbed photon current $I_{\gamma,\mathrm{abs}}$ is found from the balance of the photon currents. Incident photons can either be reflected, absorbed, or scattered and redirected to emerge back out through the surface. This assumes that all photons are reflected from the rear surface. In this balance, we neglect the emission of photons within the absorber [14].

$$(1-r)I_{\gamma,\mathrm{inc}} = I_{\gamma,\mathrm{em}} + I_{\gamma,\mathrm{abs}}$$

For the assumed isotropic and homogeneous distribution of photons with the photon current density $j_{\gamma,\Omega}$ per solid angle, the photon current absorbed in the volume $V = AL$ is

$$I_{\gamma,\mathrm{abs}} = 4\pi\alpha V j_{\gamma,\Omega}$$

The photon current leaving through the surface is

$$I_{\gamma,\mathrm{em}} = A\,(1-r)\,\frac{\pi}{n^2}j_{\gamma,\Omega}$$

The absorptance for perfect light trapping is therefore

$$a_{\mathrm{trap}} = \frac{I_{\gamma,\mathrm{abs}}}{I_{\gamma,\mathrm{inc}}} = (1-r)\frac{I_{\gamma,\mathrm{abs}}}{I_{\gamma,\mathrm{em}} + I_{\gamma,\mathrm{abs}}}$$

$$a_{\mathrm{trap}} = (1-r)\frac{4\pi\alpha L}{(1-r)\pi/n^2 + 4\pi\alpha L}$$

$$a_{\mathrm{trap}} = \frac{1-r}{(1-r)/(4n^2\alpha L)+1} \tag{7.12}$$

For small values of the absorption coefficient α, we find the absorptance $a_{\mathrm{trap}} = 4n^2\alpha L$. For a reflectance r that is not too large, it is surprising that the absorptance does not depend on the reflectance. For silicon, a_{trap} is a factor of $4n_{\mathrm{Si}}^2 \approx 50$ greater than the absorptance without light trapping. Relative to a single pass along the normal to the surface, the mean light path is then enhanced by this factor of $4n_{\mathrm{Si}}^2 \approx 50$.

Although the above derivation is based on a homogeneous distribution of photons, which is a good approximation only when $\alpha \ll 1/L$, Equation 7.12 is also valid to a high degree of accuracy for larger values of α since the saturation value $a_{\mathrm{trap}} = 1 - r$ is already reached for $\alpha < 1/L$ and cannot increase further, even for larger values of α. Figure 7.7 shows that, by light trapping, high absorptance values can be obtained even for thin silicon layers in spite of silicon's small absorption coefficient.

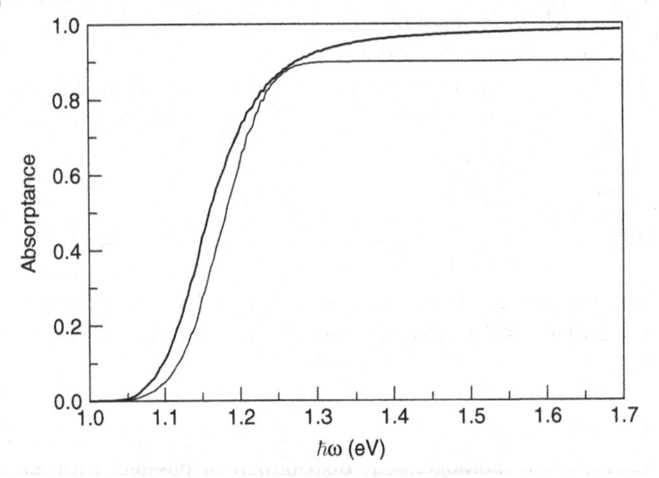

Figure 7.7 Absorptance a as a function of the photon energy $\hbar\omega$ for 20 μm thick silicon with light trapping (heavy line) and for 400 μm thick silicon without light trapping (thin line), assuming a reflectance of $r = 0.1$ in both cases.

The enhancement of the absorptance by light trapping can reach even greater values. In the derivation above, we have assumed that the photons leaving through the surface are emitted into the entire hemisphere, that is, into an effective solid angle π as a result of the Lambertian behavior of the scattering surface. Because of the reversibility of light paths, incident radiation from the entire hemisphere is able to enter the cell at the same time through the surface. For this type of light trapping, the solar cell does not have to track the Sun's position.

According to the discussion in Section 2.4.3, the maximum concentration of solar radiation is obtained when the photons emitted from the surface of an absorber are not emitted into the entire hemisphere but only toward the Sun. Theoretically, it is conceivable that with an appropriate surface structure, a photonic crystal structure, photons leaving through the surface will be directed only toward the Sun. The density of the photons in the semiconductor that are not absorbed or are only weakly absorbed then increases to a value that is larger by a factor n^2 than the photon density at the surface of the Sun. Such a structure would, however, not change the incident photon current density. It is a structure for minimum emission rather than for maximum concentration. A solar cell that only exchanges radiation with the Sun would of course have to track the position of the Sun. If we make this effort, the absorptance for absorbed photons with $\hbar\omega > \varepsilon_G$ would increase much more steeply than shown in Figure 7.7, and silicon solar cells could be made still thinner. For purely radiative recombination, a greater voltage would result, while the absorbed photon current, and with it the charge current of the solar cell, would only be slightly improved.

With the light-trapping technique, very thin solar cells can be realized with crystalline silicon, in spite of its low absorption coefficient. However, we must consider that the estimated increase in the mean light path is based on the incoherent scattering of photons, valid only when the dimensions of the surface structure and the thickness of the cell are large compared with the wavelength. For smaller structures, coherence and interference phenomena must be considered. In Figure 7.7, we see that a 20 μm thick silicon film with light trapping has a higher absorptance than a 400 μm thick wafer without light trapping. A higher absorptance results in a larger short-circuit current. In addition, the open-circuit voltage is larger for the thin cell, because impurity recombination and Auger recombination have a smaller probability in a smaller volume. The probability for radiative recombination, however, increases in comparison with the other recombination processes, since it depends only on the absorptance. Decreasing the thickness while maintaining a high absorptance makes the solar cell more ideal.

A high-efficiency, thin-film silicon solar cell would be a great achievement because of silicon's favorable properties for the environment and its chemical stability. One problem for future development is that thin silicon films require a substrate for support. The only known substrate on which single crystal silicon films can be grown is, however, crystalline silicon. Although only the crystal structure of the substrate is needed and not the purity required for a solar cell, it is still an expensive substrate. Recently, polycrystalline silicon films have been grown on a variety of substrates including glass, which is a very promising development.

Light trapping by textured surfaces also improves the properties of thick cells by reducing reflection and enhancing the absorption of photons with $\hbar\omega \approx \varepsilon_G$. The best silicon solar cell for the unconcentrated *AM*1.5 spectrum is made from very pure silicon and has all the properties mentioned above for an optimal Si solar cell [15]. Its efficiency is 24.4%. Figure 7.8 illustrates its structure.

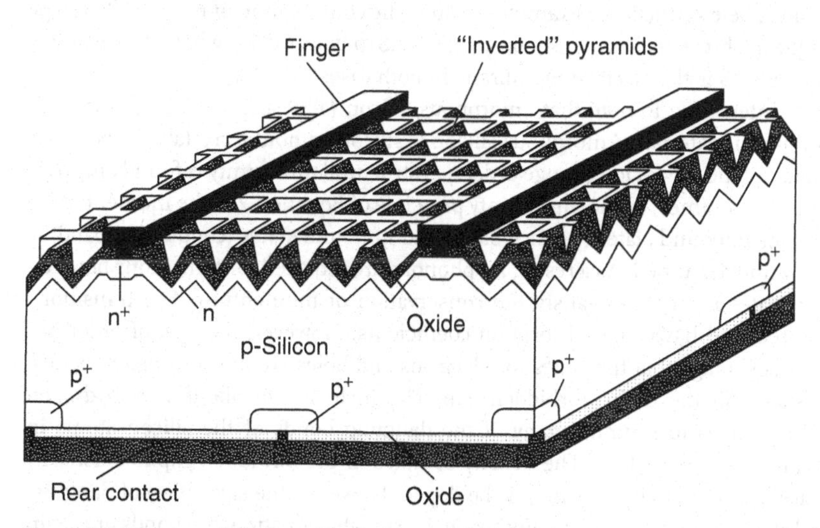

Figure 7.8 Structure of the best silicon solar cell manufactured to date with an efficiency of 24.4%, developed by M.A. Green's group at UNSW. (Courtesy of M.A. Green).

7.4
Thin-film Solar Cells

Silicon has so many advantages for solar cells that other materials can compete only when they do not share its disadvantage, the poor absorption of light. In materials competing with silicon, the transitions between the valence and conduction bands must be direct. The absorption coefficient then has a large value. For the absorption of that part of the solar spectrum that can be absorbed, a thickness of only a few micrometers is sufficient for thin-film solar cells. For the same number of recombination centers as in a thick silicon cell, a higher impurity concentration and the presence of grain boundaries in the film can be tolerated. Because of the smaller distances to the membranes at the surfaces, the diffusion lengths can also be smaller. This allows the use of materials with lower mobility. All these advantages hold the promise of significant cost reduction in the production of solar cells.

Owing to the close proximity to the surface with its high surface recombination, a well-developed membrane is required at least on the front surface. Since electron–hole pairs should not be generated in this layer, it must have a large energy gap. It is called a window layer, through which photons pass unimpeded, but which protects electrons and holes from recombining at the front contact. The interface between the window layer and the absorber should have a low density of interface states in order to prevent recombination there.

A disadvantage of many materials with direct transitions and favorable energy gaps, with the exception of amorphous silicon, is that they cannot be doped equally well n-type and p-type. The structure required for solar cells then demands heterojunctions. Examples include the combination of n-type CdS/p-type $CuInSe_2$(abbreviated as (CIS)) or n-type CdS/p-type CdTe, where the window material CdS is the electron membrane in both cases.

It is interesting to note that amorphous silicon (a-Si) also belongs to the class of thin-film materials. Amorphous silicon is silicon without a crystalline structure. Because of the lack of long-range order, i.e. structural uniformity is found only over very small volumes, by the uncertainty principle of Equation 2.4, the momentum of electrons in bound states (valence band) and unbound states (conduction band) is largely undetermined. As a result, no phonons are required for transitions between these states in order to satisfy the conservation of momentum. The transitions are direct and have large absorption coefficients. However, the lack of order has the disadvantage that the states for electrons and holes are not confined to bands; the states fill the entire forbidden gap. The inclusion of about 10% hydrogen (a-Si:H) serves to saturate many of the dangling bonds of the silicon atoms in the amorphous structure. The density of states in the forbidden gap is drastically reduced, and the material can now be doped. However, the saturation of dangling bonds with hydrogen is not totally stable. During illumination, the bonds are again broken by the capture of holes. This property, known as *the Staebler–Wronski effect*, leads to a continuous decrease of the efficiency of solar cells made of a-Si:H.

7.4.1
Minimal Thickness of a Solar Cell

The thickness of a solar cell is an important issue. It is not only that a larger amount of precious material is needed for a thicker cell but also that a thinner cell could tolerate less optimal material properties. Organic materials could be very useful for solar cells because of their good absorption properties and good luminescent quantum yields, indicating dominant radiative recombination. Their drawbacks are the very small mobilities and diffusion lengths of electrons and holes. In the common plane arrangement of an absorber between two membranes, shown in Figure 7.9(a), the diffusion lengths must be larger than the thickness of the absorber and the thickness must be larger than the penetration depth $1/\alpha$ of the photons. If this were a generally necessary condition, then many organic materials would never make good solar cells. In fact, these conditions are sufficient, but not really necessary. The necessary condition for the absorption is that there is enough absorber material to absorb the light; how the material is arranged is not important. Two separate thin layers absorb as much as a single layer that is twice as thick. The necessary condition for the transport of electrons and holes is that they must be able to reach the membranes, requiring the diffusion lengths to be larger than the distance between the membranes. The arrangement in Figure 7.9b fulfills the conditions of absorption and transport as well and would allow the use of absorber materials with arbitrarily small diffusion lengths. This principle of separating the absorber thickness from the distance between the membranes is successfully realized in the dye solar cell described in Chapter 5. There, the distance between the membranes, TiO_2 for the electrons and the I^-/I_3^--redox system for the holes, has a minimal value, the thickness of the monomolecular dye layer. Nevertheless, many dye layers provide for sufficient absorption.

The organic or plastic solar cell is another example [16]. In this solar cell, a p-type polymer serves both as an absorber and as a hole membrane. The electrons tunnel from their bound exciton state in the absorber to a mobile state in n-type fullerene molecules while the holes stay in the polymer. The polymer and the fullerene are

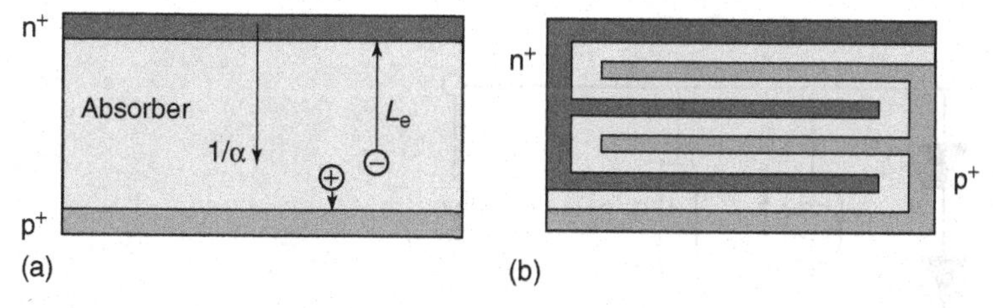

Figure 7.9 (a) In the plane arrangement of an absorber between electron and hole membranes, the diffusion lengths $L_{e,h}$ must be larger than the thickness of the absorber and the thickness must be larger than the penetration depth $1/\alpha$ of the photons. (b) Many absorbing layers in a meander-like structure combine good absorption with a small distance between the membranes.

thoroughly mixed into a blend to facilitate the charge carrier separation. This makes the arrangement different from Figure 7.9b, where care was taken that the electron membrane makes contact with only one of the electrodes and the hole membrane only with the other electrode. With the mixing of electron and hole membranes in the plastic solar cell, as well as in the dye cell, a strong inhibition of the exchange of one carrier type must be present at each of the electrodes to prevent internal shunts. In the dye cell, hole exchange between the transparent electrode on the TiO_2 and the redox system is very poor, but is excellent with the small platinum islands on the counterelectrode.

With the small distance between the membranes and the resulting large interface area, interface recombination is enhanced and may be a problem.

7.5 Equivalent Circuit

In the current–voltage characteristic for the solar cell in Equation 6.35, we can regard the current I_Q as the sum of the current through the pn-junction in the dark and the current I_{sc} from a current source, connected in parallel for the currents to add.

Figure 7.10 shows the equivalent circuit diagram, extended by two additional elements. The resistance R_p, in parallel with the two diodes of the two-diode model, represents the shunts that can occur in real solar cells across the surfaces, at pinholes in the pn-junction or at grain boundaries. The series resistance R_S accounts for all voltage drops across the transport resistances of the solar cell and its connections to a load. The current–voltage characteristic then takes the form

$$I_Q = I_{S1}\left[\exp\left(\frac{e(V - I_Q R_S)}{kT}\right) - 1\right] + I_{S2}\left[\exp\left(\frac{e(V - I_Q R_S)}{2kT}\right) - 1\right] + I_{sc} + \frac{V - I_Q R_S}{R_p} \quad (7.13)$$

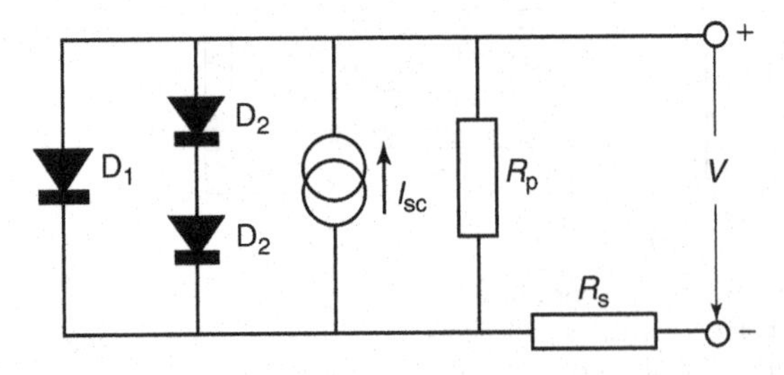

Figure 7.10 Equivalent circuit for a solar cell consisting of (from left) diode D_1 with direct recombination, diodes D_2 with impurity recombination, current source I_{sc}, parallel resistance R_p and series resistance R_S.

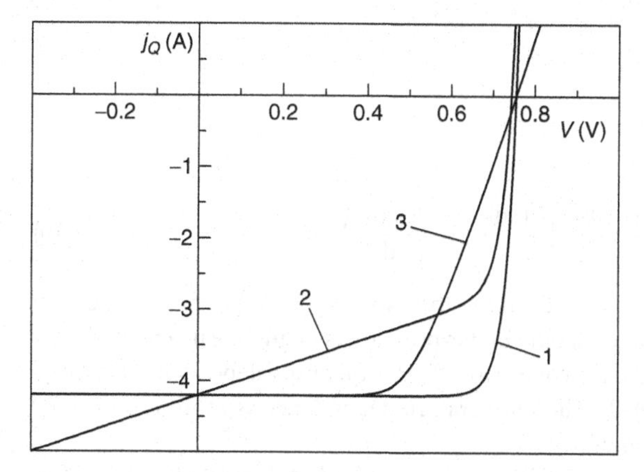

Figure 7.11 Current–voltage characteristic of a 100 cm^2 solar cell with (2) $R_S = 0\ \Omega$, $R_p = 0.5\ \Omega$; and (3) $R_S = 0.05\ \Omega$, $R_p = \infty$, compared with (1) $R_S = 0\ \Omega$ and $R_p = \infty$.

Figure 7.11 illustrates how the characteristic is affected by varying R_p and R_S separately. The effect of the series resistance alone is a displacement of the characteristic in the direction of lower voltages, in proportion to the current. The effect of the parallel resistance alone is a displacement of the characteristic to higher positive currents, in proportion to the voltage. Both effects, separately and combined, lead to a smaller fill factor *FF*.

7.6 Temperature Dependence of the Open-circuit Voltage

Solar cells deliver only a small part of the absorbed energy current as electrical energy to a load. The remainder is dissipated as heat; the solar cell must therefore have a higher temperature than the environment. For solar irradiation of 1 kWm^{-2}, the temperature difference to the environment may be some 10 K.

Heating reduces the size of the energy gap. The absorbed photon current increases, leading to a slight increase in the short-circuit current j_{sc}. The heating has a detrimental effect on the open-circuit voltage. From

$$V_{oc} = \frac{1}{e}(\eta_e + \eta_h) = \frac{kT}{e}\ln\left(\frac{n_e n_h}{n_i^2}\right) \tag{7.14}$$

we find for the temperature dependence

$$\frac{dV_{oc}}{dT} = \frac{k}{e}\ln\left(\frac{n_e n_h}{n_i^2}\right) + \frac{kT}{e}\left[\frac{1}{n_e}\frac{dn_e}{dT} + \frac{1}{n_h}\frac{dn_h}{dT} - \frac{1}{n_i^2}\frac{d\left(n_i^2\right)}{dT}\right] \tag{7.15}$$

Here

$$n_i^2 = N_C N_V \exp\left(-\frac{\varepsilon_G}{kT}\right)$$

and

$$\frac{\mathrm{d}\left(n_\mathrm{i}^2\right)}{\mathrm{d}T} = \frac{\varepsilon_\mathrm{G}}{kT^2}\, n_\mathrm{i}^2$$

It follows that

$$\frac{\mathrm{d}V_\mathrm{oc}}{\mathrm{d}T} = \frac{V_\mathrm{oc} - \varepsilon_\mathrm{G}/\mathrm{e}}{T} + \frac{kT}{\mathrm{e}}\left(\frac{1}{n_\mathrm{e}}\frac{\mathrm{d}n_\mathrm{e}}{\mathrm{d}T} + \frac{1}{n_\mathrm{h}}\frac{\mathrm{d}n_\mathrm{h}}{\mathrm{d}T}\right) \tag{7.16}$$

No general statements can be made about the expressions in parentheses, except that both are probably <0 and that the first expression is negligible in a n-conductor and the second expression in a p-conductor. The temperature dependence is caused essentially by $(V_\mathrm{oc} - \varepsilon_\mathrm{G}/\mathrm{e})/T$. The temperature dependence is more pronounced in bad cells, where V_oc is small.

For a silicon cell with $V_\mathrm{oc} = 0.6$ V and $\varepsilon_\mathrm{G} = 1.12$ eV at $T = 300$ K, $\mathrm{d}V_\mathrm{oc}/\mathrm{d}T = -1.7$ mVK^{-1}. This means that the open-circuit voltage decreases by 0.3% per degree of temperature rise. A temperature increase of 50 K reduces the open-circuit voltage by 85 mV, that is, by 14%. The efficiency is reduced accordingly.

7.7 Intensity Dependence of the Efficiency

We know from earlier discussions that maximum efficiencies are obtained for maximum concentration of the incident radiation. Since concentration means that a smaller area is required for a given energy current, it is an option for expensive solar cell materials. That an increase of the efficiency can be expected is a bonus.

The short-circuit current j_sc is simply given by the absorbed photon current and increases, as expressed by Equation 7.6, proportional to the intensity.

$$j_\mathrm{sc} = -\mathrm{e}j_{\gamma,\mathrm{abs}} \tag{7.17}$$

The open-circuit voltage V_oc defines the difference between the Fermi energies at which the total recombination rate in the cell is equal to the total generation rate given by the absorbed photon current. With $j_{\gamma,\mathrm{emit}} = j_{\gamma,\mathrm{abs}}$, we find from Equation 7.8

$$V_\mathrm{oc} = \frac{kT}{\mathrm{e}}\ln\left(1 + \frac{\mathrm{e}j_{\gamma,\mathrm{abs}}}{j_\mathrm{S}}\right) \tag{7.18}$$

where j_S is the reverse saturation current. For $j_{\gamma,\mathrm{abs}} \gg j_\mathrm{S}$, true for any observable light intensity, the open-circuit voltage V_oc increases logarithmically with the intensity.

The fill factor, finally, as given by Equation 7.5 can be approximated for $V_\mathrm{oc} \gg kT$ roughly by

$$FF = 1 - \frac{kT}{V_\mathrm{oc}} \tag{7.19}$$

and increases very slightly with the open-circuit voltage. To a first approximation, the increase of the fill factor with the intensity can be neglected.

In the efficiency

$$\eta = \frac{FF\, j_{sc}\, V_{oc}}{\int_0^\infty \hbar\omega\, \mathrm{d}j_{\gamma,\,\mathrm{Sun}}(\hbar\omega)} \tag{7.20}$$

the linear increase with the intensity of the denominator is compensated by the linear increase in the short-circuit current in the numerator, and the efficiency is seen to increase with the intensity by the logarithmic increase in the voltage. Transport resistances have not been considered in this discussion. For large currents generated by high intensities, this is certainly problematic.

7.8 Efficiencies of the Individual Energy Conversion Processes

With a theoretical limit for the efficiency of $\eta = 0.3$ for the *AM*0 spectrum, energy conversion with a solar cell is still well away from the theoretical limit of $\eta_{max} = 0.86$ for the solar heat engine of Section 2.5. It is very instructive once again to examine the processes in a solar cell individually and break down the overall efficiency into the efficiencies of the individual processes in order to recognize where the greatest losses occur. Figure 7.12 shows the individual processes schematically.

The first process is the absorption of the incident energy current. Its efficiency has to account for the photons with energy $\hbar\omega < \varepsilon_G$, which are not absorbed by the solar cell. The absorbed energy current is given by the absorbed photon current times the mean energy $\langle\hbar\omega_{abs}\rangle$ of the absorbed photons. Let us assume that each absorbed photon generates just one electron–hole pair which, at short-circuit, contributes to the current. Then

$$j_{E,\,\mathrm{abs}} = j_{\gamma,\,\mathrm{abs}}\, \langle\hbar\omega_{\mathrm{abs}}\rangle = -\frac{j_{sc}}{e}\, \langle\hbar\omega_{\mathrm{abs}}\rangle \tag{7.21}$$

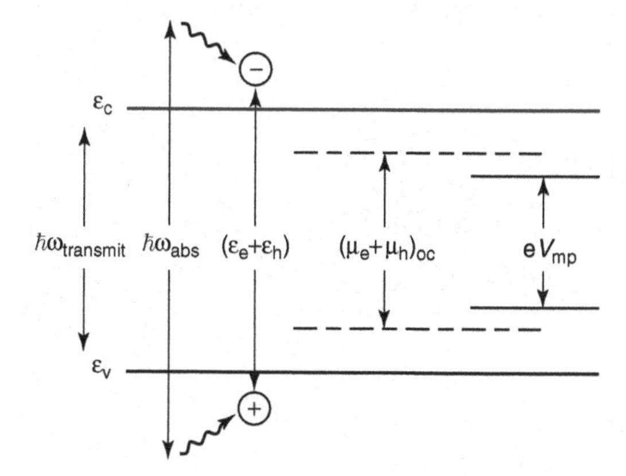

Figure 7.12 Individual processes in a solar cell.

The absorption efficiency is

$$\eta_{\mathrm{abs}} = \frac{j_{E,\mathrm{abs}}}{j_{E,\mathrm{inc}}} \tag{7.22}$$

The second process is the thermalization of the electron–hole pairs, each produced initially with the mean energy $\langle \hbar\omega_{\mathrm{abs}} \rangle$, which by thermalization is reduced to $\langle \varepsilon_e + \varepsilon_h \rangle = \varepsilon_G + 3kT$. The efficiency of this process is

$$\eta_{\mathrm{thermalization}} = \frac{\langle \varepsilon_e + \varepsilon_h \rangle}{\langle \hbar\omega_{\mathrm{abs}} \rangle} \tag{7.23}$$

The third factor defining the maximum chemical energy $(\mu_e + \mu_h)_{\mathrm{oc}} = eV_{\mathrm{oc}}$ that can be obtained from the energy $\langle \varepsilon_e + \varepsilon_h \rangle$ of the electron–hole pairs is the efficiency accounting for the difference between free energy and energy with a thermodynamic factor

$$\eta_{\mathrm{thermodynamic}} = \frac{eV_{\mathrm{oc}}}{\langle \varepsilon_e + \varepsilon_h \rangle} \tag{7.24}$$

This maximum chemical energy per electron–hole pair at open-circuit is, however, completely lost with the emitted photons. To gain energy, we have to go to the maximum power point. This brings in the last factor, the fill factor FF, which determines how much of the maximum chemical energy current $-j_{\mathrm{sc}}\, V_{\mathrm{oc}}$ the solar cell delivers at the maximum power point as an electrical energy current $-j_{\mathrm{mp}}\, V_{\mathrm{mp}}$.

$$FF = \frac{j_{\mathrm{mp}} V_{\mathrm{mp}}}{j_{\mathrm{sc}} V_{\mathrm{oc}}} \tag{7.25}$$

The product of all these efficiencies gives the overall efficiency

$$\eta = \underbrace{\frac{j_{E,\mathrm{abs}}}{j_{E,\mathrm{in}}}}_{\eta_{\mathrm{abs}}} \underbrace{\frac{\langle \varepsilon_e + \varepsilon_h \rangle}{\langle \hbar\omega_{\mathrm{abs}} \rangle}}_{\eta_{\mathrm{thermalization}}} \underbrace{\frac{eV_{\mathrm{oc}}}{\langle \varepsilon_e + \varepsilon_h \rangle}}_{\eta_{\mathrm{thermodynamic}}} \underbrace{\frac{j_{\mathrm{mp}} V_{\mathrm{mp}}}{j_{\mathrm{sc}} V_{\mathrm{oc}}}}_{FF} = \frac{-j_{\mathrm{mp}} V_{\mathrm{mp}}}{j_{E,\mathrm{in}}} \tag{7.26}$$

For silicon, and in particular, for the 20 μm thick cell with light trapping, whose absorptance is shown in Figure 7.7, exposure to the $AM1.5$ spectrum gives the following values:

$$\langle \hbar\omega_{\mathrm{abs}} \rangle = 1.80\ \mathrm{eV}$$

$$\langle \varepsilon_e + \varepsilon_h \rangle = \varepsilon_G + 3kT = 1.2\ \mathrm{eV}$$

$$j_{\mathrm{sc}} = -413\ \mathrm{Am^{-2}} \qquad j_{\mathrm{mp}} = -401\ \mathrm{Am^{-2}}$$

$$V_{\mathrm{oc}} = 0.770\ \mathrm{V} \qquad V_{\mathrm{mp}} = 0.702\ \mathrm{V}$$

The efficiencies are therefore

$$\eta_{\mathrm{abs}} = 0.74$$

$$\eta_{\mathrm{thermalization}} = 0.67$$

$$\eta_{\text{thermodynamic}} = 0.64$$

$$FF = 0.89$$

The overall efficiency is then $\eta = 0.74 \times 0.67 \times 0.64 \times 0.89 = 0.28$.

The efficiencies for thermalization and for the conversion of the energy of the electron–hole pairs into chemical energy are particularly small and thus in need of improvement.

7.9 Problems

7.1 The current–voltage characteristic of a pn-junction shall be described by Equation 6.27 with $j_S = 10^{-12}$ mA cm^{-2} and $j_{sc} = -35$ mA cm^{-2}. Recalculate the characteristic adding

(a) a series resistance normalized with respect to area $\tilde{R}_s = 10\ \Omega\text{cm}^2$

(b) a normalized parallel resistance of $\tilde{R}_p = 50\ \Omega\text{cm}^2$

(c) both.

8
Concepts for Improving the Efficiency of Solar Cells

We saw in the previous chapter that even avoiding all nonradiative recombination processes leaves us with a solar cell efficiency well below the theoretical maximum value of $\eta = 0.86$, derived in Chapter 2 as the upper limit for solar energy conversion. The main reasons were identified as losses by thermalization and the nonabsorption of low-energy photons. In order to improve the efficiency, we must focus primarily on reducing these losses. We now discuss different methods by which this can be accomplished, in principle. The underlying conditions are idealized, often to such an extent that it is difficult to imagine how they can be met, in practice. Nevertheless, it is still important to examine these methods in order to recognize the principles for possible improvements. A detailed discussion of these methods was recently presented by Green [17].

8.1
Tandem Cells

The reduction of thermalization losses and the improvement in the absorption efficiency can be simultaneously achieved by offering the solar cell only photons within the narrow energy interval $\varepsilon_G < \hbar\omega < \varepsilon_G + d\varepsilon$ and processing the other photons by solar cells with a different band gap. Cells operated in this way are known as *tandem cells*.

For a black-body solar spectrum, a solar cell with the energy gap ε_G and the idealized absorptance $a(\hbar\omega < \varepsilon_G) = 0$, $a(\hbar\omega \geq \varepsilon_G) = 1$ has the short-circuit current

$$j_{sc} = -e\frac{\Omega_S}{4\pi^3\hbar^3c^2}\int_{\varepsilon_G}^{\infty}\frac{(\hbar\omega)^2}{\exp\left(\hbar\omega/kT_S\right)-1}\,d\hbar\omega \tag{8.1}$$

Figure 8.1 gives the short-circuit current j_{sc} as a function of the energy gap ε_G/e. Following thermalization, the energy current flowing into the electron–hole pairs is

$$j_{E,eh} = \frac{-j_{sc}\,\varepsilon_G}{e} = j_{\gamma,abs}\,\varepsilon_G \tag{8.2}$$

Using $\langle\varepsilon_e + \varepsilon_h\rangle = \varepsilon_G$ instead of $\varepsilon_G + 3kT$ here, we divide the overall efficiency η somewhat differently (and not quite correctly) into the thermalization and the thermodynamic efficiencies.

Physics of Solar Cells: From Basic Principles to Advanced Concepts. Peter Würfel

ISBN: 978-3-527-40857-3

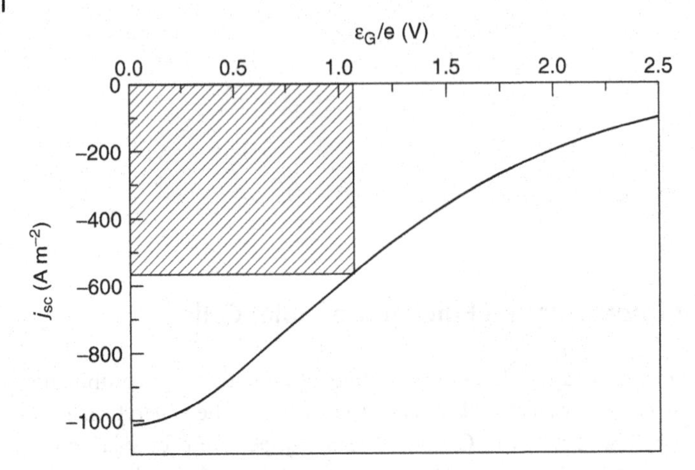

Figure 8.1 Short-circuit current of a solar cell as a function of the energy gap ε_G of its absorber for a black-body spectrum with $T_S = 5800$ K.

The shaded rectangle in Figure 8.1 indicates the energy current $j_{E,\mathrm{eh}}$ transferred to the electron–hole pairs after thermalization, for the energy gap at which $j_{E,\mathrm{eh}}$ has its maximum value.

The area under the curve $j_{\mathrm{sc}}(\varepsilon_G/\mathrm{e})$ is the entire incident energy current $j_{E,\mathrm{Sun}}$ coming from the Sun. This can best be seen if we determine the variation of $\mathrm{d}j_{\mathrm{sc}}$ of the short-circuit current for a small variation $\mathrm{d}\varepsilon_G$ in the band gap of the absorber

$$\mathrm{d}j_{\mathrm{sc}} = -\mathrm{e}\frac{\Omega_S}{4\pi^3\hbar^3c^2}\frac{\varepsilon_G^2}{\exp\left(\varepsilon_G/kT_S\right) - 1}\,\mathrm{d}\varepsilon_G \tag{8.3}$$

and then integrate over the variation in the absorbed energy current $-(\varepsilon_G/\mathrm{e})\,\mathrm{d}j_{\mathrm{sc}}$, caused by the variation in the band gap $\mathrm{d}\varepsilon_G$; thus,

$$j_E = -\int_0^\infty \frac{\varepsilon_G}{\mathrm{e}}\,\mathrm{d}j_{\mathrm{sc}} = \frac{\Omega_S}{4\pi^3\hbar^3c^2}\int_0^\infty \frac{\varepsilon_G^3}{\exp\left(\varepsilon_G/kT_S\right) - 1}\,\mathrm{d}\varepsilon_G \tag{8.4}$$

Since the value of a definite integral does not depend on the names of the variables (we could also call it $\hbar\omega$ instead of ε_G), Equation 8.4 is recognized to describe the density of the incident solar energy current. The largest rectangle, shaded in Figure 8.1, showing the energy current transferred to the electron–hole pairs after thermalization, corresponds to 42% of the incident energy current. The area under the curve to the right of the rectangle gives the energy current, which is lost by thermalization. The area below the rectangle down to the $j_{\mathrm{sc}}(\varepsilon_G/\mathrm{e})$ curve is the energy current, which is not utilized, because it is not absorbed by the solar cell. A solar cell with $\varepsilon_G = 1.1$ eV would thus have an efficiency of 42%, if all of the energy of the electron–hole pairs could be converted into electrical energy, i.e. if the thermodynamic efficiency were equal to 1. This value, called the *ultimate efficiency* by Shockley and Queisser [18] can, however, not be achieved at room

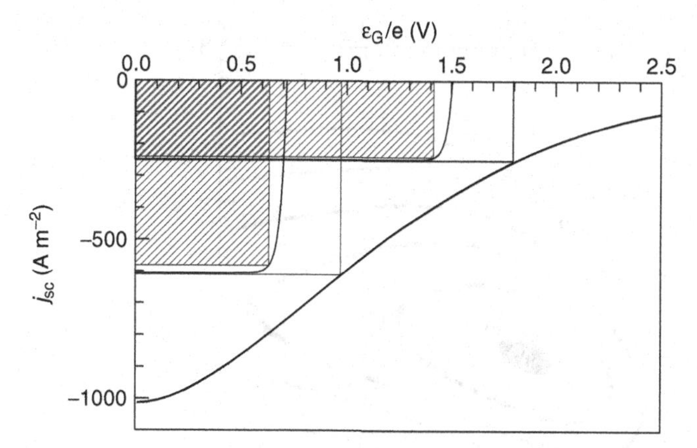

Figure 8.2 Current–voltage characteristics of two solar cells with energy gaps $\varepsilon_{G1} = 1.8$ eV and $\varepsilon_{G2} = 0.98$ eV.

temperature, because entropy must not be annihilated and some of the energy of the electron–hole pairs is room temperature heat.

Figure 8.2 shows the same curve as Figure 8.1, this time illustrating how two solar cells with different energy gaps ε_{G1} and ε_{G2} divide the incident energy current of the *AM*0 spectrum. The energy current first falls on the cell with the greater energy gap ε_{G1}, which absorbs all photons with $\hbar\omega \geq \varepsilon_{G1}$ and transmits all photons with $\hbar\omega < \varepsilon_{G1}$. The cell behind, with the lower energy gap, then absorbs the photons with $\varepsilon_{G2} \leq \hbar\omega < \varepsilon_{G1}$.

For the two cells depicted in Figure 8.2 the current–voltage characteristics are also shown together with the cross-hatched rectangles indicating the maximum electrical energy current that the cells deliver. Here again, we assume that only radiative recombination is present.

Figure 8.3 gives the efficiency obtained from two cells having the energy gaps ε_{G1} and ε_{G2} when their energy currents are added. For the *AM*0 spectrum, the optimal combination consists of $\varepsilon_{G2} = 1.0$ eV and $\varepsilon_{G1} = 1.9$ eV, yielding an overall efficiency of $\eta = 0.44$.

If we extend the division of the incident spectrum onto an infinite number of solar cells optically in series, with continuously decreasing energy gaps, the entire energy current from the Sun will then be transferred to energy of the electron–hole pairs, and no thermalization losses will occur at all. With this configuration, each cell absorbs those photons from the incident photon current that have energies in the interval $\varepsilon_G \leq \hbar\omega < \varepsilon_G + d\varepsilon_G$, assuming that the energy gaps of adjoining cells differ by $d\varepsilon_G$.

Under open-circuit conditions, with radiative recombination only, exactly the same number of photons is emitted as is absorbed, which is always true for radiative recombination. This time, however, the emitted spectrum is identical with the absorbed spectrum, since there are no thermalization losses. All emitted photons have the temperature of the cell, but they have nonzero chemical potentials that depend on their energy. If we also assume maximum concentration of the

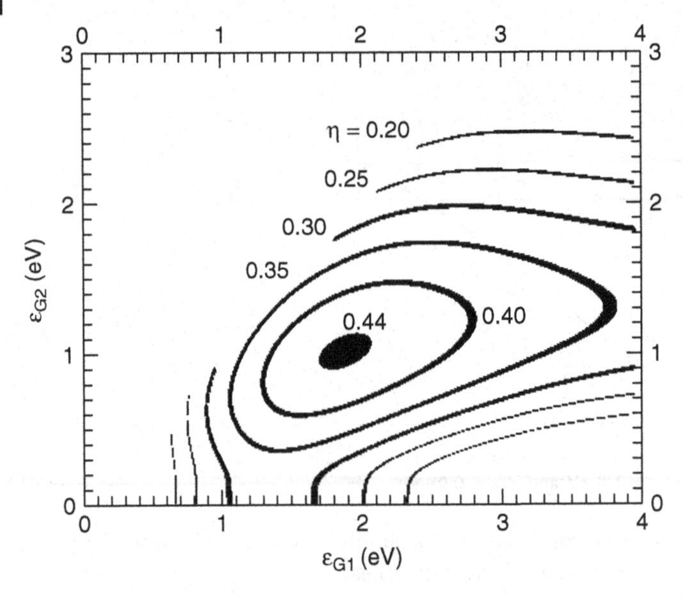

Figure 8.3 Efficiency for two solar cells in tandem operation, with energy gaps ε_{G1} and ε_{G2} for the *AM0* spectrum when their energy currents are added.

incident solar radiation, in accordance with the generalized Planck radiation law (3.102), the chemical potential μ_γ of the emitted photons for one cell that absorbs and emits photons of energy $\hbar\omega = \varepsilon_G$ is then

$$\mu_\gamma = \hbar\omega\left(1 - \frac{T_0}{T_S}\right) = \varepsilon_G\left(1 - \frac{T_0}{T_S}\right) \tag{8.5}$$

In our treatment of maximum efficiencies in Chapter 2 we saw that the Landsberg efficiency requires an absorber capable of absorbing solar radiation without creation of entropy, although its temperature is that of the environment and not that of the Sun. The infinite tandem is one such absorber. However, in contrast to the infinite tandem, in which the production of entropy is prevented by the production of chemical energy, the Landsberg efficiency also requires that the absorber emits photons with $\mu_\gamma = 0$. The infinite tandem can only prevent the production of entropy in the open-circuit state, in which the absorbed energy current is completely reemitted back to the Sun, while the Landsberg efficiency process assumes that no entropy is created at maximum power.

At maximum power of the tandem stack of solar cells, at which each cell of the infinite tandem operates at a different voltage V_{mp} and emits photons with a different chemical potential $\mu_\gamma = eV_{mp}$, but always at the same temperature T_0, the emission spectrum is identical with the emission spectrum of an infinite tandem of monochromatic thermal absorbers feeding energy into Carnot engines, as was discussed in Section 2.5. The upper limit for the efficiency that can be obtained with an infinite tandem is $\eta = 0.86$ and is identical with that for photothermal conversion discussed in Section 2.5.

8.1.1
The Electrical Interconnection of Tandem Cells

For the optimal absorption of the incident photons, the cells are arranged one after the other, optically in series. Electrical contacts between the cells, which absorb photons, must be avoided. This allows only for connection of the cells electrically in series. The voltages of the different cells must all have the same sign. This requires that the same type of membrane of all the cells, e.g. the p-type membrane, must be facing the Sun. It follows that the n-type membrane of the preceding cell is facing the p-type membrane of the following cell. The problem now is that holes flow out of the p-membrane and electrons out of the n-membrane. The transfer of the charge at the interface between these membranes requires that the holes recombine with the electrons. The difference between the Fermi energies, normally required for the recombination, however, would give rise to a voltage opposite in sign to the voltages of the solar cells. Figure 8.4 shows two cells that are series connected in such a way that electrons flowing out of the n-membrane of the left cell and holes flowing out of the p-membrane of the right cell recombine in a tunnel junction. In the tunnel junction, electrons and holes belong to the same Fermi distribution ($\varepsilon_{F2} = \varepsilon_{F3}$) and recombination proceeds without a difference in the Fermi energies. As is seen in the figure, the Fermi energy must lie in the conduction band of the n-membrane on one side and in the valence band of the p-membrane on the other side, requiring high doping concentrations ($n_D, n_A > N_C, N_V$). Since the tunnel junction is very thin, the absorption by its high free-carrier concentrations is very small.

Thermalization was found to be necessary for the conversion of solar heat into chemical energy in Chapter 2. Thermalization losses are reduced or even prevented when the generated electrons and holes populate only narrow energy ranges at the

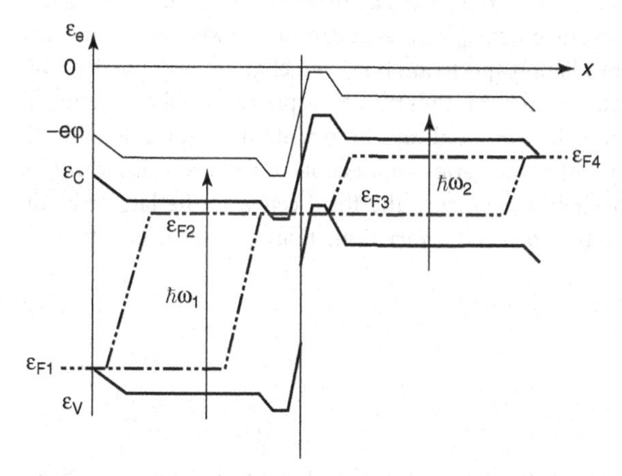

Figure 8.4 Electrical series connection of two solar cells by a tunnel junction provides efficient recombination of electrons and holes without requiring a difference of their Fermi energies.

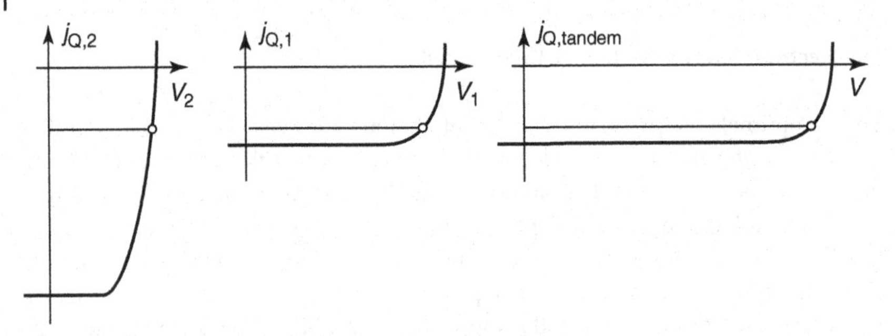

Figure 8.5 Current–voltage characteristic for the series connection of two solar cells with different short-circuit currents and open-circuit voltages. For each value of the current j_Q the voltages V_1 and V_2 are added to give the total voltage V.

band edges, as explained above. This condition can be expressed in a different way. Prevention of thermalization losses requires many different Fermi distributions (one for each energy range) with many different Fermi energies. The tandem in Figure 8.4 has three different Fermi energies, one more than for a single-material solar cell.

Series connection forces the same charge current to flow through all the cells. Figure 8.5 shows how the current–voltage characteristic of a tandem results from the characteristics of two cells with different short-circuit currents and different open-circuit voltages. For a given current, the voltages V_1 and V_2 are determined from the characteristics of the individual cells and added to give the voltage V of the tandem configuration. The cell with the smaller short-circuit current determines the total current.

To prevent losses due to a series connection, as in Figure 8.5, the energy gaps must be chosen so that the currents j_{mp} at the maximum power points are the same for all cells. The solar radiation spectrum, however, changes over the day and the year due to different path lengths through the atmosphere, and the equality of the currents j_{mp} of different cells cannot always be maintained. For a tandem of three cells, a configuration where the series connection of the two cells with the smaller band gaps is connected in parallel to the third cell with the largest band gap is found to be less sensitive to spectral variations than a series connection of all three cells [19].

8.2
Concentrator Cells

In Chapter 2, we considered how strongly we can focus the incident solar radiation. For concentrated radiation, the same power is incident onto a solar cell with smaller area than for nonconcentrated radiation. Another advantage is that concentrated radiation can be processed with greater efficiency, as discussed in Chapter 7.

In areas with much direct, unscattered solar radiation, the additional expense for concentration is rewarded by a better efficiency from a smaller solar cell. When the radiation is concentrated with lenses or mirrors, the solar cell sees only a part of the hemisphere and, in the limiting case of maximum concentration, only the Sun. The greater the concentration, the more carefully the concentrating system must track the path of the Sun. The concentration of radiation also has disadvantages. The improvement in the efficiency assumes that, in spite of an increase in the incident radiation, the temperature of the cell remains the same. But in fact the temperature of the cell rises, and with poor cooling the efficiency can even decrease with increasing concentration. Another disadvantage is the result of the larger electrical currents causing larger voltage losses across the series resistance of the cell and in the leads.

Special solar cells known as *concentrator cells* have been developed for concentrated radiation. Because of the higher temperatures, semiconductors with a larger energy gap are always advantageous. However, they must be able to absorb a sufficiently large part of the solar spectrum. Cells made of GaAs are well suited for this. Because of their smaller surface areas, more expensive materials and more expensive constructions become economical. For concentrator systems, tandem cells based on III–V compounds have become cost effective. For silicon concentrator cells, a more complicated design was developed, where the n- and p-type membranes for the electrons and the holes are in the form of point contacts, all placed on the back side of the cell as shown in Figure 8.6.

The electrons and holes flow out of the n- and p-doped regions, placed alternately on the rear side. This is different from the structures discussed up to now and has the advantage that the contacts do not cause shadowing and can be kept relatively large to avoid series resistances. To reduce Auger recombination to a minimum, the greatest part of the solar cell is not doped. The highly doped regions required as membranes and to conserve the entropy per particle are limited to the contacts. For concentration by a factor of 100 an efficiency of 28% was obtained with a silicon cell. This was, however, for a temperature of 25 °C, which requires intensive cooling [20]. For a tandem, a record efficiency of 40.7% for a monolithic

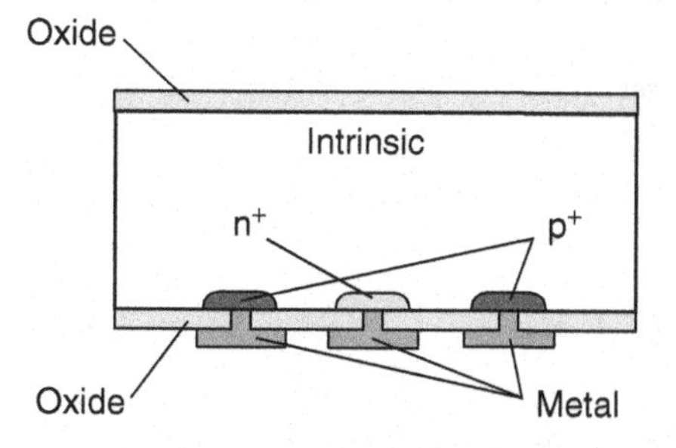

Figure 8.6 The p- and n-type membranes for holes and electrons and the metal contacts of the point-contact cell for concentrated radiation are both placed on the back side of the cell.

configuration of three cells consisting of GaInP, GaInAs, and Ge operating at a concentration of the solar radiation by a factor of 240 was reported [21].

8.3 Thermophotovoltaic Energy Conversion

The solar–thermal conversion method of Section 2.5 can be modified to be applicable to solar cells. Figure 8.7 illustrates the principle.

A focusing optical system is used to concentrate the solar radiation onto an intermediate absorber that, as a result, is heated to the temperature T_A. Solar cells with an energy gap ε_G are placed concentrically around the intermediate absorber. They have an interference filter on their surface, which transmits all photons with $\varepsilon_G \leq \hbar\omega \leq \varepsilon_G + d\varepsilon$ without loss and reflects all other photons, which cannot be used optimally, back to the intermediate absorber. These photons together with those emitted by the solar cell and transmitted by the filter help maintain the temperature T_A of the intermediate absorber. If the recombination is entirely radiative and the photons emitted by the cells are reabsorbed by the absorber and are therefore not lost, the cells can be operated close to their open-circuit voltage and thus have the efficiency derived in Equation 4.5 for the conversion of the absorbed photon energy, i.e.

$$\eta_{cell} = 1 - \frac{T_0}{T_A} \tag{8.6}$$

Since this is the Carnot efficiency, which was used for the efficiency of the heat engine of the solar–thermal conversion process described in Section 2.5, we find for the thermophotovoltaic conversion process the same efficiency as for the solar–thermal conversion; thus,

$$\eta = \left[1 - \left(\frac{T_A}{T_S}\right)^4\right]\left(1 - \frac{T_0}{T_A}\right) \tag{8.7}$$

with a maximum value of $\eta = 0.85$ at an absorber temperature of $T_A = 2478$ K.

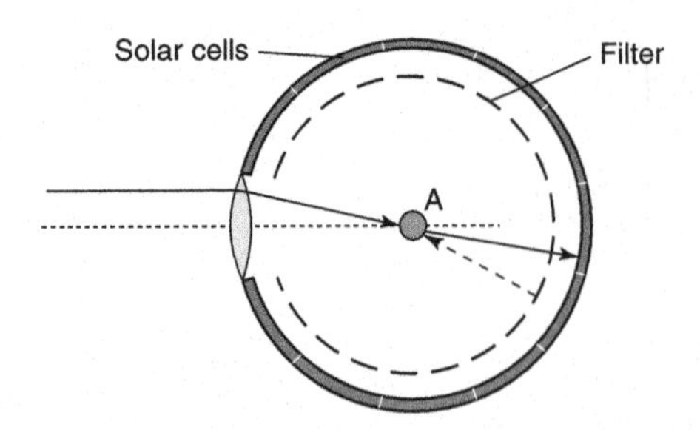

Figure 8.7 In the thermophotovoltaic converter, the intermediate absorber is surrounded in an evacuated cavity by solar cells illuminated by its thermal radiation.

The implementation of this concept in practice is difficult for two reasons. At the optimal temperature of the intermediate absorber of $T_A = 2478$ K, all materials evaporate so strongly that the interference filter is quickly covered with an opaque layer. Moreover, in practice it is not possible to construct an interference filter transmitting only in a narrow energy interval and reflecting the rest of the spectrum while also being free of absorption. With the use of, e.g. silicon solar cells, only a very small portion of the photons emitted from the intermediate absorber have the required energy $\hbar\omega \geq \varepsilon_G$. Even very little absorption by the interference filter of all the other photons leads to a considerable loss. Smaller band gap materials like GaSb are more favorable for thermophotovoltaic conversion.

In the construction principle of Figure 8.7, another drawback is hidden. The absorber must be able to emit almost as much energy as it absorbs in the narrow energy interval transmitted by the filter. Since T_A is much smaller than the temperature of the Sun T_S, the emitting area must be much larger than the absorbing area. A factor of 4, as provided by the arrangement in Figure 8.7 is far from sufficient.

This problem can be solved by an even wilder idea called *thermo-photonics* [22]. Emission of only photons with $\hbar\omega \geq \varepsilon_G$ can be accomplished by placing a semiconductor on the emitting area of the intermediate absorber. (To prevent transmission by the semiconductor of smaller-energy photons emitted by the absorber, a mirror must be placed between the semiconductor and the absorber.) When the semiconductor is supplied with membranes such as a solar cell, it can be operated as a light-emitting diode (LED). A LED is the same engine as a solar cell, only operated in reverse, just as a refrigerator or heat pump is a reversely operated heat engine. If some of the power delivered by the solar cells is used to drive the LED on the absorber, the emitted intensity is enhanced enormously. Although some of the energy emitted by the LED is free energy supplied by the solar cells, most of it is heat supplied by the absorber. As a result, the area and the temperature of the emitter can be reduced, provided an LED can be made that works at about 1000 °C with close to 100% external quantum efficiency.

With the arrangement shown in Figure 8.7, however, the intermediate absorber does not have to be heated by the Sun. It is also possible to heat the intermediate absorber in another way, e.g. by burning gas. Radiation losses to the environment would then not have to occur, because the cavity can be completely closed, optically. The conversion of heat into electrical energy in this manner was first proposed in the Soviet Union for nuclear reactors. The basic concept was to surround incandescent reactor fuel elements with solar cells. Fortunately, no one had the courage to try this out.

8.4 Impact Ionization

Electrons and holes possessing large kinetic energies as a result of generation by high-energy photons can dissipate their kinetic energy in two ways. One is by elastic

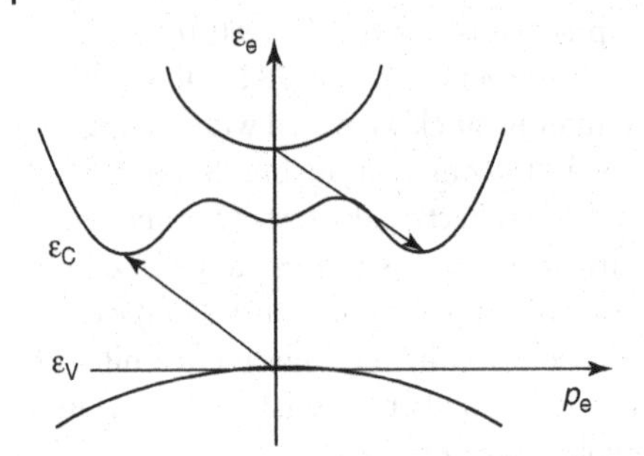

Figure 8.8 Transition of an electron from a higher band to the minimum of the conduction band by impact ionization in an indirect semiconductor, resulting in the additional generation of an electron and a hole at the band edges.

collisions with the lattice atoms, in which energy is transferred in small portions to the lattice atoms until thermal equilibrium with the lattice is established. The other is by inelastic collisions with the lattice atoms in which, by impact ionization, another electron is knocked off its chemical bond or, in other words, in which a free electron and a free hole are produced, as shown in Figure 8.8. Both processes take place in parallel and compete with each other. With elastic collisions, the excitation of lattice vibrations is at the expense of the energy of the electron–hole system, while the number of the electrons and holes remains constant. With impact ionization, the absorbed energy remains in the electron–hole system, but is more uniformly distributed over a larger number of electrons and holes than were originally generated by the absorption of the photons. Impact ionization, therefore, looks very promising for solar energy conversion because some of the energy removed from the electrons and holes during thermalization is used to generate additional electron–hole pairs [23].

To examine the efficiency of the impact ionization process, we will exclude the competing process, namely, the interaction of electrons and holes with the lattice vibrations, which leads to thermalization at constant concentrations. The electrons and holes are then isolated from the lattice vibrations. They do not "know" about the temperature of the lattice and cannot come into thermal equilibrium with the lattice. Collisions between electrons and holes are, however, permitted. This ensures that electrons and holes have a uniform temperature T_A, although this is not the same as the lattice temperature. Finally, according to the principle of detailed balance, we must expressly consider Auger recombination as the inverse process of impact ionization [24].

We will now examine the temperature and electrochemical potentials of the electrons and holes under these conditions. The simplest answer, unfortunately, may be the most difficult to understand. It is based on the difference between thermalization and impact ionization: whereas during thermalization by scattering with phonons, no electrons (or holes) are annihilated or created, so that their number remains constant, during impact ionization and its inverse process, Auger recombination, the number of electrons and holes changes. This has significant consequences for the values of the electrochemical potentials of the electrons and holes. The change of the particle numbers by impact ionization and Auger recombination is unrestricted, except for the condition to establish a minimum of the free energy of the electrons and the holes. Thus,

$$dF = \cdots + \eta_e \, dN_e + \eta_h \, dN_h + \cdots = 0$$

With impact ionization and Auger recombination, electrons and holes are always created or annihilated in pairs, that is, $dN_e = dN_h = dN$ and

$$dF = \cdots + (\eta_e + \eta_h) \, dN + \cdots = 0$$

Because the number of particles does not remain constant and is not tied to other particle numbers as in a chemical reaction, $dN \neq 0$, so that

$$\eta_e + \eta_h = 0$$

We will attempt to make this result more plausible. Let us assume that the free energy of the electrons and holes describes a state with $\eta_e + \eta_h > 0$. With a reduction in the number of particles due to Auger recombination, i.e. with $dN < 0$ and therefore $dF < 0$, the free energy can be further reduced. With the reduction in the number of particles, $\eta_e + \eta_h$ also decreases, until for $\eta_e + \eta_h = 0$ a further reduction of the particle number no longer reduces the free energy and equilibrium is established between impact ionization and Auger recombination. Since the total energy of the electron–hole system is preserved, by reducing the particle number, Auger recombination leads to an increase in the energy per particle, i.e. in the mean kinetic energy of the electrons and holes. This in turn leads to an increase in their temperature and in the number of electrons and holes capable of participating in impact ionization, until the rate of impact ionization is exactly the same as the rate of Auger recombination and $\eta_e + \eta_h = 0$.

If, however, scattering with the lattice atoms and Auger recombination and impact ionization all occur at high rates, the temperature of the electrons and holes will be the same as the lattice temperature, and in equilibrium with impact ionization and Auger recombination, the only possible state is one with $T_{eh} = T_0$ and $\eta_e + \eta_h = 0$, which does not permit the conversion of energy. It is therefore very important that thermalization and impact ionization, together with its inverse process, do not occur simultaneously and with similar probabilities. It would not improve the efficiency of solar cells, but in fact reduce it, if the probability for impact ionization (and Auger recombination, inevitably) were slightly increased while the interaction with phonons predominates.

We thus establish the fact that the interaction with the lattice vibrations maintains the temperature of the electron–hole system constant at the lattice temperature T_0 and, on exposure to light, produces a state with $\eta_e + \eta_h > 0$. Impact ionization and Auger recombination, on the other hand, maintain a state with no separation of the Fermi energies, $\eta_e + \eta_h = 0$, but in the absence of interaction with the lattice vibrations and on exposure to light produce a state with $T > T_0$.

The problem of how to obtain electrical energy from hot electrons and holes then remains to be solved. Energy conversion by means of impact ionization first produces hot electrons and holes with no chemical energy. Chemical energy, and finally electrical energy, must be obtained in subsequent steps.

8.4.1 Hot Electrons from Impact Ionization

The temperature T_A of the electrons and holes in the absorber can be found very easily from the magnitude of the emitted energy current. In contrast to the interaction with phonons, where at open circuit the emitted *photon* current is equal to the absorbed photon current, for impact ionization at open circuit the emitted *energy* current must be equal to the absorbed energy current, since by impact ionization/Auger recombination the electron–hole system does not lose energy. The chemical potential of the emitted photons is then $\mu_\gamma = \eta_e + \eta_h = 0$, and we can calculate the temperature T_A of the electrons and holes by using Planck's law in Equation 2.34 for the emitted energy current. For maximum concentration of the incident solar radiation, we find, of course, $T_A = T_S$ at open circuit.

8.4.2 Energy Conversion with Hot Electrons and Holes

For conventional solar cells based on thermalization of electrons and holes in the absorber, complete conversion of chemical energy into electrical energy was achieved by membranes, n-type for the transport of electrons to one contact and p-type for the transport of holes to the other contact. This type of membrane is not sufficient for hot carriers. In addition to the selective transport of electrons and holes, the membranes must now also serve the thermodynamic function of producing chemical energy from the heat of the electrons and holes by cooling them down to the temperature of the environment.

We will discuss this problem for the electrons; the solution can then easily be applied to the holes as well. If we would allow an exchange of electrons between absorber and membrane for all electron energies in the absorber, the thermalization of the electrons in the membrane would lead to a large energy loss, from $(3/2)kT_A$ in the absorber to $(3/2)kT_0$ in the membrane (less, however, than in a conventional solar cell). Secondly, with the unimpeded exchange of electrons between the absorber and the membrane, the electrons in the absorber would be cooled as well and would no longer be capable of impact ionization. This would result in a state with $\eta_e + \eta_h = 0$ and $T_A = T_0$. However, as we have seen earlier, the entire energy loss caused by thermalization can be prevented if the electrons in the membrane only occupy states over a narrow range $\Delta\varepsilon_e$ at the energy ε_e, as shown in Figure 8.9.

For $\Delta\varepsilon_e < kT_0$, the occupation of the electron states in the membrane cannot significantly change by interaction with the phonons. As a result, the entropy of the electrons also remains unchanged and thermalization takes place isentropically. Since the number of particles remains constant during thermalization in the membrane, the electrochemical potential of the electrons increases. Figure 8.9 demonstrates that the same process takes place with the holes at an energy ε_h in

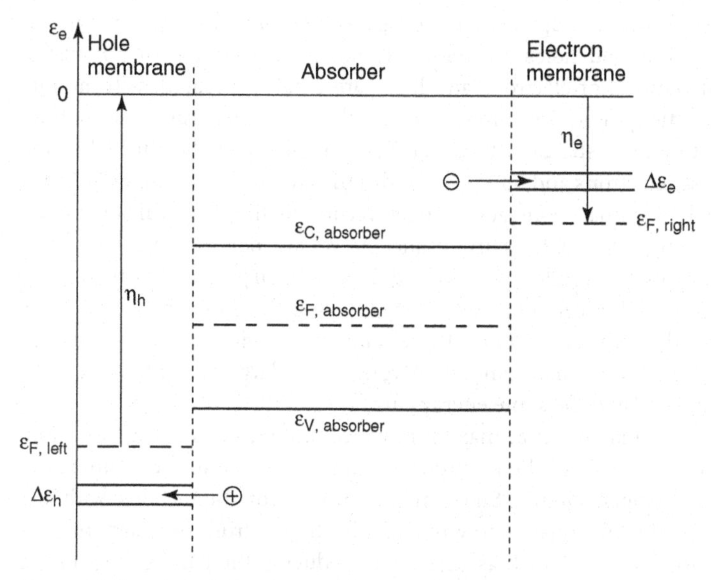

Figure 8.9 Energies of electrons and holes in the absorber, in which impact ionization and Auger recombination are in equilibrium at $T_A > T_0$, and in membranes through which electrons and holes flow outward and where they are in temperature equilibrium with the environment at $T = T_0$.

the hole membrane. The isentropic cooling therefore produces the chemical and electrochemical energy per electron–hole pair,

$$\mu_e + \mu_h = \eta_e + \eta_h = (\varepsilon_e + \varepsilon_h)\left(1 - \frac{T_0}{T_A}\right) \tag{8.8}$$

The arrangement of Figure 8.9 is a working solar cell. The voltage is

$$V = \frac{(\eta_e + \eta_h)}{e} \tag{8.9}$$

and the current is

$$j_Q = e\,\frac{(j_{E,\,\mathrm{absorbed}} - j_{E,\,\mathrm{emitted}})}{(\varepsilon_e + \varepsilon_h)} \tag{8.10}$$

We can visualize its operation by increasing the current from zero, the open-circuit situation, where the emitted energy current equals the absorbed energy current. With increasing charge current, the energy in the electron–hole system decreases and with it the electron–hole temperature T_A in the absorber. Owing to the lower temperature, the emitted energy current is reduced and so is the voltage. The current rises until, at zero voltage, the short-circuit situation is obtained. A still larger current may be withdrawn by applying a negative voltage, spending energy from a battery, which would cool the remaining electrons and holes down to $T_A < T_0$.

It is interesting that the open-circuit voltage is determined by the energies at which the electrons and holes are removed and not by the absorber material. If the rates of removal of electrons and holes are small compared with the rate of impact ionization/Auger recombination and that of carrier–carrier scattering, which we assume to be the case, their equilibrium will hardly be affected by the removal process. Electrons and holes removed at the energies ε_e and ε_h will quickly be replenished by impact ionization/Auger recombination and carrier–carrier scattering. Although the charge current and the voltage depend on the energies with which electrons and holes are removed, it is very surprising that the energy current delivered by the cell, obtained by multiplying Equations 8.9 and 8.10, is independent of the removal energies. Large removal energies give a large voltage and a small current and small removal energies give a large current and a small voltage, both resulting in the same energy current.

The efficiency is maximum at maximum concentration of the solar radiation, when the temperature T_A of the electrons and holes is equal to the temperature T_S of the Sun at open circuit. Since in the absence of interactions with the lattice vibrations the absorbed energy remains in the electron–hole system, it is advantageous to absorb as much as possible by reducing the band gap ε_G of the absorber material to zero. The electron–hole system is then a black body and, according to Equation 2.25, absorbs the energy current σT_S^4 and emits the energy current σT_A^4 at the temperature T_A. With Equations 8.9 and 8.10, the efficiency with which electrical energy is delivered is

$$\eta = \frac{j_Q V}{\sigma T_S^4} = \frac{\sigma(T_S^4 - T_A^4)}{\sigma T_S^4}\left(1 - \frac{T_0}{T_A}\right) = \left(1 - \frac{T_A^4}{T_S^4}\right)\left(1 - \frac{T_0}{T_A}\right) \tag{8.11}$$

and is therefore identical with the efficiency of the ideal solar heat engine in Equation 2.53 in Section 2.5 and that of the thermophotovoltaic conversion process discussed in the previous section. The efficiency has its maximum value of $\eta_{max} = 0.85$ at a temperature of the electron–hole system of $T_A = 2478$ K if we assume a temperature of $T_S = 5800$ K for the Sun.

Figure 8.10 shows that the efficiency at full concentration falls off with increasing energy gap, because of the decreasing absorption. Without concentration, for $\Omega = \Omega_S$, that is, for the *AM0* spectrum, however, a nonzero energy gap is preferable, otherwise more photons would be emitted than absorbed at small photon energies. For $\varepsilon_G > 0$ the balance becomes more favorable.

An earlier proposal for a hot-carrier solar cell by Ross and Nozik [25] that did not account for impact ionization and Auger recombination, finds even higher efficiencies for narrow-gap semiconductors under less than full concentration of the solar radiation. Similar to a conventional solar cell, the process of carrier–carrier scattering, while leading to a uniform temperature, is assumed to leave the number of electron–hole pairs unchanged, increasing by one for each absorbed photon and decreasing by one for each emitted photon. This assumption leads to high temperatures and negative chemical potentials of the electron–hole pairs for less than full concentration. Since electrons and holes are withdrawn through mono-energy contacts, scattering of the carriers with each other is necessary to replenish

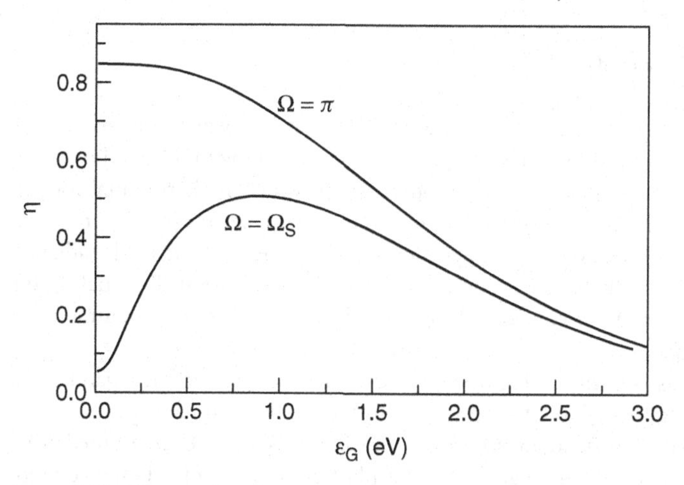

Figure 8.10 Efficiency for a hot-carrier cell with impact ionization for nonconcentrated incident solar radiation with $\Omega = \Omega_S$ and for maximum concentration with $\Omega = \pi$.

the carriers in the energy range from where they are withdrawn. A problem is that a distinction between scattering events that keep the carrier concentrations constant and impact ionization and Auger recombination which do not is physically impossible in narrow-gap semiconductors. The problem becomes obvious for the case where electron–hole pairs are withdrawn with an energy that is smaller than the average energy of the absorbed photons. The more electron–hole pairs are withdrawn, the more energy per pair piles up for the remaining electron–hole pairs. As a result, their temperature increases beyond any reasonable limit, far beyond the Sun's temperature. No such problems are encountered when impact ionization and Auger recombination are taken into account.

We thus see that impact ionization and Auger recombination allow ideal energy conversion, provided that interaction with the lattice vibrations is excluded. However, no material in bulk form is known in which these conditions are even approximately fulfilled. In a (very) thin film, however, one can imagine that electrons and holes can be removed in much less than 10^{-12} s, long before they are thermalized.

8.5
Two-step Excitation in Three-level Systems

If interaction with phonons cannot be prevented, thermalization losses can be reduced by dividing the incident spectrum over more than one transition as we have seen with tandem cells. In a three-level system, where the levels can be bands as well, three different transitions may occur in a single material: directly from the lower level to the upper level and in addition by a two-step process from the lower level to the intermediate level and from there to the upper level. In both ways, electrons are generated at the upper level and holes at the lower level.

8.5.1
Impurity Photovoltaic Effect

In Section 3.6.2, we have discussed nonradiative transitions between the bands and an impurity level. Impurities with energies for electrons in the middle of the energy gap were found to greatly enhance recombination, which is detrimental for the efficiency. In the analysis, generation of electron–hole pairs by optical transitions was neglected. Now we do just the opposite. Our model now permits only radiative transitions between the bands and to and from an impurity level. Thermalization of free charge carriers is considered, but not impact ionization or nonradiative recombination [26].

The model, as shown in Figure 8.11, has states in the valence band with $\varepsilon_e \leq \varepsilon_V$, at the impurity level with ε_{imp}, and in the conduction band with $\varepsilon_e \geq \varepsilon_C$. To optimally use the incident spectrum, the photons will be distributed over the different transitions in such a way that photons capable of a higher energy transition, e.g. band–band, are not wasted in lower energy transitions. For the impurity energy ε_{imp} in the lower half of the energy gap (not in the middle), photons having energies $\varepsilon_{imp} - \varepsilon_V \leq \hbar\omega < \varepsilon_C - \varepsilon_{imp}$ are exclusively absorbed in transitions from the valence band to the impurity. Photons having $\varepsilon_C - \varepsilon_{imp} \leq \hbar\omega < \varepsilon_C - \varepsilon_V$ are exclusively absorbed in transitions from the impurity to the conduction band, and photons having $\hbar\omega \geq \varepsilon_C - \varepsilon_V$ provide for the band–band transitions.

The absorption properties of the impurities are characterized by optical cross sections, $\sigma_{V,i}$ for transitions from the valence band to the impurity and $\sigma_{i,C}$ for transitions from the impurity to the conduction band. Optical cross sections are of the same order of magnitude as geometrical cross sections, 10^{-15} cm^2 being a typical value. Although optical cross sections vary with energy, we assume them to be constant over the energy range of absorbable photons.

The electrons and holes are assumed to have high mobilities ressulting in their homogeneous distribution, even though they are generated inhomogeneously. The steady-state concentrations belonging to a given value of the charge current follow

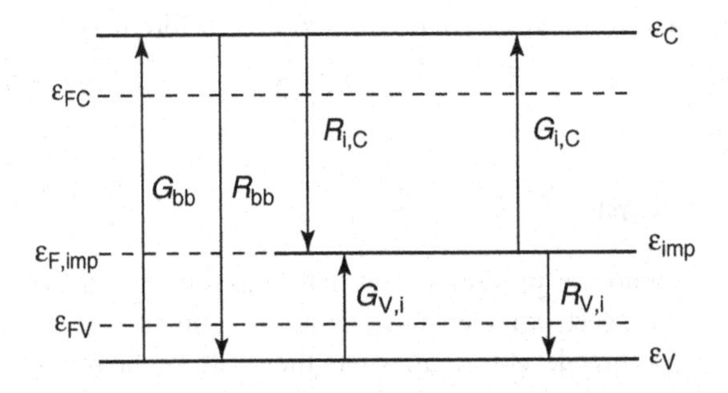

Figure 8.11 In addition to radiative band–band transitions with the rates G_{bb} and R_{bb}, radiative transitions between the bands and the impurity are taken into account. Nonradiative transitions are excluded.

from the continuity equations for the particle densities, in which in addition to generation and recombination we consider the contribution to the charge current by the divergence of the electron and hole currents. From the concentrations of electrons and holes, the sum of their electrochemical potentials and thus the voltage can be derived. The continuity equations are

$$\frac{\partial n_e}{\partial t} = G_{bb} + G_{i,C} - R_{bb} - R_{i,C} - \operatorname{div} j_e = 0 \tag{8.12}$$

$$\frac{\partial n_h}{\partial t} = G_{bb} + G_{V,i} - R_{bb} - R_{V,i} - \operatorname{div} j_h = 0 \tag{8.13}$$

$$\frac{\partial n_{e,imp}}{\partial t} = R_{i,C} - G_{i,C} + G_{V,i} - R_{V,i} = 0. \tag{8.14}$$

In the last equation, the divergence of the particle current is missing since electrons in the impurities are considered immobile, so they do not contribute to the current.

Since these three equations are not independent, we need the charge neutrality as an additional equation, as in Section 3.6.2. Because of the high absorption required, however, the impurity concentration n_{imp} is now no longer negligible compared with the densities of the electrons and holes. For this reason, most impurities must be electrically neutral, that is, either occupied, if they are donorlike, or unoccupied, if they are acceptor like. Both situations are unfavorable for the desired impurity absorption, since for transitions from the valence band to the impurity they must be unoccupied and for transitions from the impurity to the conduction band they must be occupied. The smaller of the two transition rates will determine the rate at which electrons are excited to the conduction band and holes to the valence band by two-step impurity transitions. For optimal absorption properties, we therefore choose half of the impurities to be donorlike and half acceptor like, and

$$\rho_Q = e\left(n_h - n_e + \frac{n_{imp}}{2} - n_{e,imp}\right) = 0 \tag{8.15}$$

Alternatively, if the impurities are donorlike, an occupation probability of 1/2 is achieved by doping the material additionally with half as many shallow acceptors as there are impurities. Either way, the Fermi energy in the dark has to coincide with the impurity level.

The generation rates, averaged over the thickness d, are given by

$$G_{bb} = \frac{1}{d}\int_{\varepsilon_C-\varepsilon_V}^{\infty} a_{bb}\, \mathrm{d}j_\gamma(\hbar\omega)$$

$$G_{i,C} = \frac{1}{d}\int_{\varepsilon_C-\varepsilon_{imp}}^{\varepsilon_C-\varepsilon_V} a_{i,C}\, \mathrm{d}j_\gamma(\hbar\omega) \tag{8.16}$$

$$G_{V,i} = \frac{1}{d}\int_{\varepsilon_{imp}-\varepsilon_V}^{\varepsilon_C-\varepsilon_{imp}} a_{V,i}\, \mathrm{d}j_\gamma(\hbar\omega)$$

The absorptance for the band–band transitions is assumed to be $a_{\mathrm{bb}} = 1$, whereas the absorptance for the impurity transitions depends on the concentration and occupation of the impurities,

$$a_{\mathrm{i,C}} = 1 - \exp(-\alpha_{\mathrm{i,C}}\, d); \quad a_{\mathrm{V,i}} = 1 - \exp(-\alpha_{\mathrm{V,i}}\, d)$$

where the absorption coefficients follow from Equation 3.97

$$\alpha_{\mathrm{i,C}} = \sigma_{\mathrm{i,C}}\, n_{\mathrm{imp}}\, (f_{\mathrm{i}} - f_{\mathrm{C}}) \quad \text{and} \quad \alpha_{\mathrm{V,i}} = \sigma_{\mathrm{V,i}}\, n_{\mathrm{imp}}\, (f_{\mathrm{V}} - f_{\mathrm{i}})$$

These equations determine the generation rates not only due to illumination but also in the dark state with its incident 300 K background radiation. According to the principle of detailed balance, in this state of chemical equilibrium with the background radiation, the recombination rates must have the same value as the generation rates. The principle of detailed balance is accounted for if we write the recombination rates in terms of the generalized Planck radiation law. The rate of downward transitions per energy from a level j to a level i averaged over the thickness d is, from Equation 3.102,

$$\frac{\mathrm{d}R_{i,j}}{\mathrm{d}\hbar\omega} = \frac{1}{d}\, a_{i,j}\, \frac{1}{4\pi^2\hbar^3 c_0^2}\, \frac{(\hbar\omega)^2}{\exp\{[\hbar\omega - (\varepsilon_{\mathrm{F},j} - \varepsilon_{\mathrm{F},i})]/kT\} - 1} \tag{8.17}$$

This rate is then integrated over the energy range associated with the band to band, valence band to impurity, and impurity to conduction band transitions to give the actual recombination rates.

The charge current j_Q finally results from the integral of the divergence of the electron current (or the hole current) over the thickness d of the cell and, because of the assumed homogeneous distribution of the electrons and holes, it is given by

$$j_Q = -e\, \mathrm{div}\, j_{\mathrm{e}}\, d = -e\, \mathrm{div}\, j_{\mathrm{h}}\, d \tag{8.18}$$

For a given charge current, the continuity equations 8.12–8.14 are solved for the positions of the Fermi energies subject to charge neutrality. This yields the current–voltage characteristic of the cell, since the voltage is given by $V = (\varepsilon_{\mathrm{FC}} - \varepsilon_{\mathrm{FV}})/e$. From its maximum power point the efficiency is finally determined.

The result for different energy gaps and optimized positions of the impurity level can be seen in Figure 8.12 for the *AM*0 spectrum. For this calculation, a high impurity concentration was assumed ensuring $a_{\mathrm{i,C}} = a_{\mathrm{V,i}} = 1$. The equations outlined above, however, allow us to account for smaller concentrations and inadequate occupation of the impurities, which may even vary in the course of the current–voltage characteristic.

The efficiency reaches a maximum value of $\eta = 0.46$ for an energy gap $\varepsilon_{\mathrm{C}} - \varepsilon_{\mathrm{V}} = 2.4$ eV and an impurity level at $\varepsilon_{\mathrm{imp}} - \varepsilon_{\mathrm{V}} = 0.93$ eV.

This result reminds us of the improvement of the efficiency by tandem cells because, here as well, the incident spectrum is divided over different transitions, leading to smaller thermalization losses. Figure 8.13 shows that the occupation of the various states is represented by more than two, namely, three, different Fermi energies, a condition for reduced thermalization losses known from the

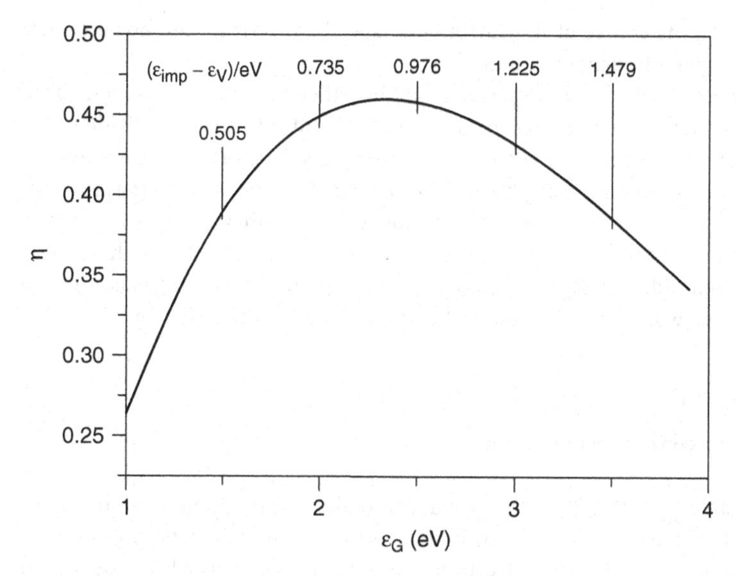

Figure 8.12 Efficiency as a function of the energy gap $\varepsilon_C - \varepsilon_V$ for radiative band–band transitions and radiative transitions between the bands and an impurity level at ε_{imp}. Nonradiative transitions are excluded. The numbers at the curve give the optimal position of the impurity level with regard to the valence band for selected band gaps.

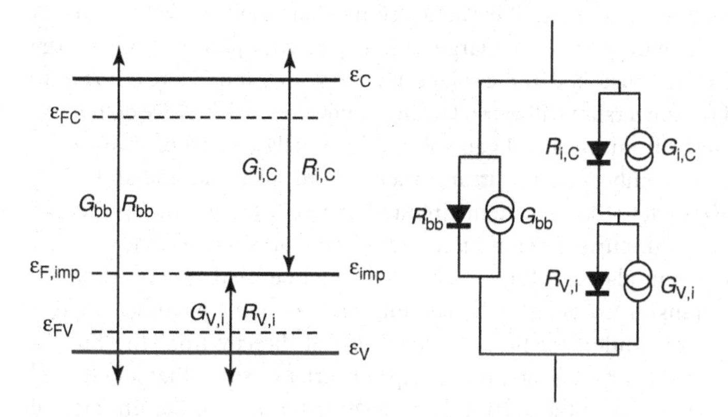

Figure 8.13 Equivalent circuit for a solar cell with an impurity level between valence and conduction bands as shown in Figure 8.11.

discussion of tandem cells. In fact, since the rate of recombination transitions is determined by the difference between the Fermi energies for the states involved, each transition can be represented by a current–voltage characteristic. This leads to the equivalent circuit shown in Figure 8.13. The solar cell representing the band–band transition is connected in parallel to a series connection of two solar cells representing the transitions involving the impurity level. Variations of the absorptance in the course of the current–voltage characteristic can, however, not be treated in an equivalent circuit model.

As for tandem cells, it is expected that the efficiency increases when more than one impurity level is present and the incident spectrum is divided into smaller portions over more transitions. Ensuring good absorption properties for all transitions, however, is a problem. Moreover, it must be emphasized that nonradiative recombination has been excluded. Although the optimal position of the impurity level for optical transitions is not in the middle of the energy gap, where nonradiative recombination is most probable, including nonradiative recombination will certainly reduce the improvement expected from impurity transitions.

8.5.2
Up- and Down-conversion of Photons

A considerable loss of energy in a solar cell is due to the photons with energy $\hbar\omega < \varepsilon_G$, which are not absorbed. It would be very convenient if two or more of these useless photons could be converted into one photon with energy $\hbar\omega \geq \varepsilon_G$, which could then be absorbed by the solar cell. In the following discussion, ε_G defines the band gap of the solar cell for which small-energy photons will be up-converted. That such an up-conversion of the photon energy is not forbidden by thermodynamics is demonstrated by Figure 8.14, which shows a device consisting of a tandem of two small band gap solar cells connected to an LED. These solar cells absorb small-energy photons, and due to their series connection, deliver a voltage that is large enough to drive the LED with a large band gap to emit photons with energy $\hbar\omega \geq \varepsilon_G$, which can be absorbed by the solar cell we have in mind. There is no doubt that this type of up-conversion will work. One may, however, ask why we do not use the electrical energy from the small-gap solar cells directly instead of investing it into an LED. We remember that the arrangement of two solar cells and an LED in Figure 8.14 is identical to the equivalent circuit of a three-level system represented by two bands and an impurity level in Figure 8.13 in the previous section.

A three-level system placed behind a solar cell could be used to convert small-energy photons transmitted by the solar cell into higher-energy photons supplied to the solar cell, in addition to the photons absorbed directly from the Sun as shown in Figure 8.15. A mirror behind the up-converter ensures that all emitted photons are directed toward the solar cell. A closer inspection of the impractical but functioning up-converter in Figure 8.14 reveals that it is better represented by a four-level system than by a three-level system. The sum of the band gaps of the two solar cells is larger than the band gap of the LED. In fact, detailed calculations as outlined in the last section show that an energy loss, indicated in Figure 8.15 at the upper level, is necessary to prevent the recombination of the electron–hole pairs via the intermediate level with the reemission of two small-energy photons [27].

Figure 8.16 shows a substantial improvement in the efficiency of a solar cell for an incident 6000 K black-body spectrum. As always, the efficiency is larger for maximum concentration than for nonconcentrated radiation. The possible

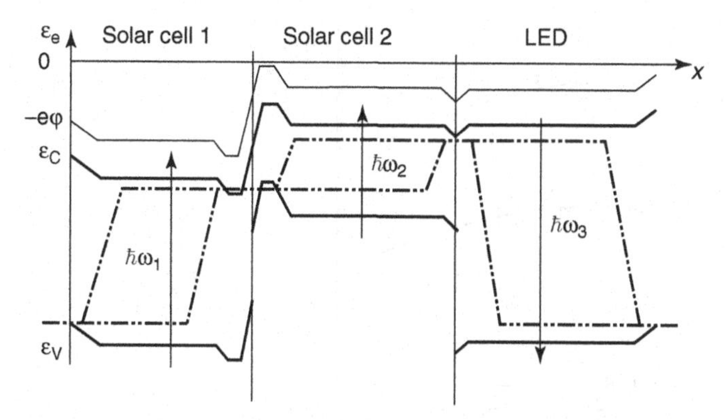

Figure 8.14 Two solar cells with small band gaps drive an LED with a large band gap, to emit photons useful for a large band gap solar cell, thereby up-converting two small-energy photons into one higher-energy photon.

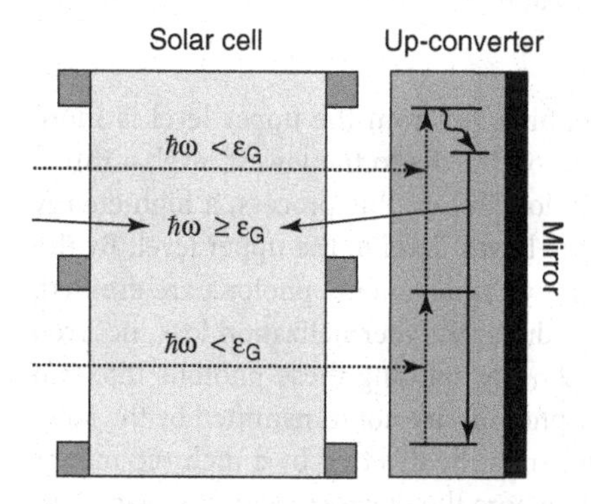

Figure 8.15 An up-converter behind a solar cell absorbs small-energy photons, transmitted by the solar cell, in a two-step excitation process. Higher-energy photons with $\hbar\omega \geq \varepsilon_G$ emitted by the up-converter generate additional electron–hole pairs in the solar cell.

efficiencies of a solar cell with up-conversion and of the impurity photovoltaic effect are very similar. Both use two-step excitations to absorb otherwise nonabsorbed photons. The up-conversion, however, has distinct advantages. First, the up-converter is a purely optical device and can consist of a material such as an organic dye, in which electrons and holes are virtually immobile, but which has a high quantum efficiency. Second, the up-converter is a separate device that can be applied to existing well-developed bifacial solar cells. Third, since the up-converter is separated from the solar cell, it would interfere very little with the recombination processes in the solar cell. Applying an up-converter to a solar cell would not do any harm but could only improve the solar cell's efficiency, even if it is not working quite as well as theoretically predicted.

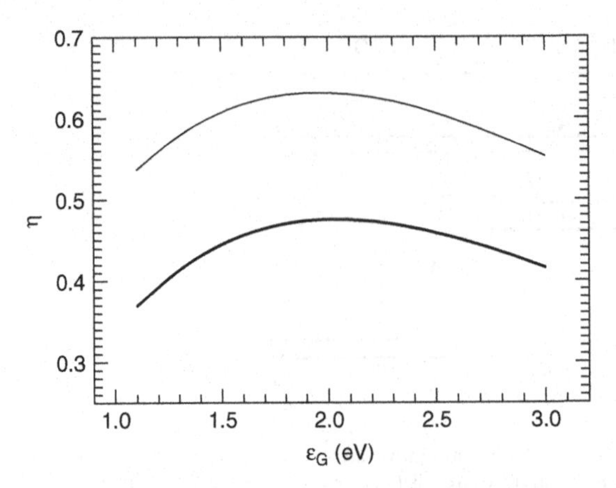

Figure 8.16 Efficiency of a solar cell as a function of its band gap ε_G operating with directly absorbed and with up-converted photons from a 6000 K black-body spectrum, from a solid angle $\Omega_S = 6.8 \times 10^{-5}$ (thick line) and with maximum concentration from $\Omega = \pi$ (thin line).

Since in a three-level system the recombination from the upper level is more probable via the intermediate level instead of directly to the lowest level, a three-level system can be used for down-conversion [28]. In this process, a high-energy photon is absorbed in a transition from the lowest level to the upper level. By the back transition, via the intermediate level, two small-energy photons are emitted. Applied to a solar cell, a down-converter reduces the thermalization loss incurred by the absorption of photons with $\hbar\omega > 2\,\varepsilon_G$ by splitting these photons into two photons with $\hbar\omega \geq \varepsilon_G$. Since high-energy photons are not transmitted by the solar cell, the high-energy part of the spectrum must be diverted by a dichroic mirror or other means to the rear side of the cell, where the down-converter is placed. In addition, another dichroic mirror transmitting high-energy photons but reflecting small-energy photons is applied to the back of the down-converter, where it prevents the loss of small-energy photons produced in the down-converter. The efficiency for a solar cell combined with a down-converter on its backside is shown in Figure 8.17 by the thin line as a function of the band gap of the solar cell.

A much simpler and more elegant method would be to place the down-converter on the front side of the solar cell. In this arrangement, all the incident solar photons, which the solar cell could absorb, are absorbed by the down-converter. Photons with $\varepsilon_G \leq \hbar\omega < 2\,\varepsilon_G$ are absorbed in transitions involving the intermediate level and larger-energy photons cause direct transitions from the lower to the upper level. Nearly all photons emitted by the down-converter have an energy $\hbar\omega > \varepsilon_G$ and could be absorbed by the solar cell. However, being on the front side of the solar cell, no mirror directing all the emitted photons into the solar cell can be applied and one might think that this deficiency leads to the loss of one half of the photons being emitted through the front surface toward the Sun. This is not necessarily the case. We remember that the probability for emission is proportional to the

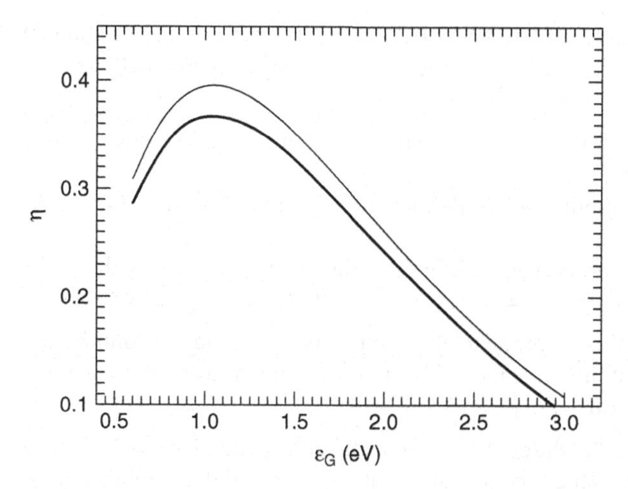

Figure 8.17 Efficiency of a solar cell as a function of its band gap ε_G operating with directly absorbed and with down-converted photons from a nonconcentrated 6000 K black-body spectrum for a down-converter placed on the rear side of the solar cell (thin line). The efficiency for operation only with photons from the down-converter when it is placed on the front side is smaller (thick line).

density of photon states. This makes the emitted photon currents proportional to the square of the index of refraction. If a material for the down-converter is chosen that has the same large index of refraction as the solar cell, all the photons emitted toward the solar cell enter the solar cell without reflection, whereas most of the photons emitted toward the front surface are totally reflected.

For the down-converter on the rear side of the solar cell, a maximum efficiency of almost 40% is found. A smaller efficiency of 36% is found, caused by the loss of down-converted photons, when the down-converter is placed on the front side of the solar cell. An index of refraction of $n = 3.6$ was used for the down-converter and the solar cell, which prevents the loss by total internal reflection of most down-converted photons.

Small band gap solar cells are advantageous in combination with a down-converter, whereas higher band gap solar cells are favorable with an up-converter. For both systems, the calculations consider only radiative transitions, which may, however, be closer to reality in materials that do not require good transport properties for electrons or holes.

8.6 Problems

8.1 List the different steps that take place in a semiconductor device that lead to the conversion of solar heat radiation into electrical energy. Comment on the phenomena that cause losses and make suggestions on how to overcome these if possible.

8.2 Consider a sample of n-doped bulk Si with a donor concentration of $n_D = 10^{15}\ \mathrm{cm^{-3}}$, fully ionized. Regard three different recombination mechanisms:

- Radiative recombination as defined in Equation 3.55 with $B = 3 \times 10^{-15}\ \mathrm{cm^3\,s^{-1}}$.
- Auger recombination as defined in Equation 3.63 with $C_e = C_h = 10^{-30}\ \mathrm{cm^6\,s^{-1}}$.
- Impurity recombination (via acceptor-like impurity states) with $n_{imp} = 10^{13}\ \mathrm{cm^{-3}}$, $\varepsilon_{imp} - \varepsilon_V = 0.7\ \mathrm{eV}$, $\tau_{min,e} = 10^{-3}\ \mathrm{s}$, and $\tau_{min,h} = 10^{-5}\ \mathrm{s}$.

(a) Calculate and plot the total recombination rate and the contribution of each of the three recombination mechanisms as a function of the minority carrier density n_h.

(b) Calculate the difference of the quasi-Fermi energies $\varepsilon_{FC} - \varepsilon_{FV}$ as a function of the recombination rate. Consider the recombination as being caused by each of the three mechanisms alone and by the combination of all of them, respectively.

Prospects for the Future

In the first chapter, we saw that our present energy economy cannot continue in its current form, because we are in the process of changing the environment in which we live. Mankind and all other forms of life have developed over millions of years by adapting to the surrounding conditions in a continuous process of evolution. We have good reason to describe changes to these conditions, which are too rapid for life to adapt to, as natural disasters. Life on Earth is an extremely complex system, and even today we do not fully understand the interrelations within this system. If we do not wish to risk life as a whole or parts thereof by significant departures from the present equilibrium, we should tolerate only minor changes in the prevailing conditions. Every system reacts linearly to minor changes, even our complex ecosystem. Small changes to the natural processes result in small changes of the properties of the environment. The restriction to minor changes means that we may make use only of processes existing in the natural state of equilibrium. The burning of wood, coal, oil, and gas occurs as a natural process as well, resulting in the natural production of CO_2, CO, and SO_2. Burning of wood, coal, oil, and gas could thus be tolerated, if their burning by man only caused minor changes to natural processes. However, we are presently proceeding rapidly to violate this condition. The amounts produced by mankind are no longer minor changes. The condition of causing only minor changes is, in any case, violated when processes are used or substances produced that do not exist in nature and it is almost impossible to predict how the environment will react. This is particularly true for many waste products from nuclear energy use. An example that shows what can happen if formerly nonexisting substances are introduced into the environment is that of chlorinated fluoro-hydrocarbons (CFC), which do not occur at all in nature. They were regarded as entirely harmless, since they are nontoxic and chemically inert. It came as a great surprise to learn that they, in fact, destroy the ozone layer and also are, to a large extent, responsible for the greenhouse effect. Many more examples exist, where the violation of the condition of minor changes has led to unpleasant surprises.

No such surprises are to be expected when electricity is generated from solar energy by solar cells. In making use of the processes taking place in a solar cell, we are only linking ourselves into processes that would, in any case, occur without

Physics of Solar Cells: From Basic Principles to Advanced Concepts. Peter Würfel

ISBN: 978-3-527-40857-3

us. In our absence, solar radiation would be absorbed by the Earth and, in part, be reflected back into space. In the course of this process, the Earth heats up to just the temperature at which it can reemit the energy current absorbed from the Sun. It is very important that we do not alter this process significantly. Viewed from the standpoint of thermodynamics, the solar heat, very valuable at the Sun's temperature of roughly 6000 K, is cooled down to the Earth's temperature, where it is practically worthless and is then emitted into space. What is changed, if the solar radiation is processed by solar cells? Part of the absorbed heat (in fact, most of it in real systems with efficiencies of 20% or less) is cooled at the site of the solar cells down to the temperature of the environment in the same way as it would be without the solar cells. The electrical power generated by the solar cells is then rerouted through consumers before friction and other dissipative processes finally degrade it to heat at the temperature of the environment, from where it is emitted into space. The energy balance between the absorbed and the emitted energy currents remains unchanged. With the use of solar cells, we simply allow the natural process of cooling the solar heat to take place in ways of greater benefit to us.

The preceding chapters have not only shown that solar cells are well suited for obtaining electrical power from solar energy, they have also shown that there is no better way to do this than with the intelligent utilization of solar cells, e.g. in tandem cell arrangements, because the possible efficiencies of these systems coincide with general efficiency limits predicted by thermodynamics. This has two consequences. On the one hand, there is no need to continue searching for other fundamentally more efficient methods of harnessing solar energy. On the other hand, since we cannot hope for future discoveries and technologies with significantly greater efficiencies, there is no reason to wait any longer to begin the serious development of a solar-energy economy.

Our present energy economy consumes oxygen and produces CO_2. Because of the fast and extensive spreading of gases in the atmosphere, this is a global, and not simply a local, problem. For heavily populated and industrialized areas such as Germany, where so much oxygen is burnt that none would be left to breathe, this spreading of the gases is very fortunate. But it also makes it very difficult for politicians to decide on radical changes in the present energy economy, because continuing with the present energy consumption has little consequences locally, or, the other way round, the required very considerable local efforts are not rewarded on a local basis if they are not implemented globally.

A global energy supply by solar energy on the present level must be easily possible, otherwise we would already be suffering from substantial global warming. If our power requirements were not small compared with the solar-energy current reaching the Earth, their coverage from resources would result in an increase in the temperature of the Earth, even without the greenhouse effect in order to allow for the emission of this additional energy into space.

A quick estimate indicates that a solar-based global energy economy could, in principle, be implemented relatively easily. The greater part of the globally consumed 10×10^{13} kWh/a, is used to produce low-temperature heat, for heating buildings and for cooking. Even in countries with less than average sunshine like Sweden, most of these requirements can be met using well-insulated solar warm-water collectors. The rest, roughly 5×10^{13} kWh/a, could be generated by solar cells. Most of this would be used for the production of hydrogen, since this is an easily transported and easily stored form of chemical energy. In sunny areas, with incident solar radiation of more than 2000 kWh/(a m^2), a total efficiency of no more than 10% over an area of 500 km $\times$ 500 km equal to 2.5×10^{11} m^2, would be sufficient. Much larger areas are available in the sunny deserts. Nevertheless, covering such an enormous area with solar cells is presently unimaginable. The problems with a future solar-based energy economy would be alleviated, if we could reduce our energy requirements, or at least maintain them at their present level.

This vision of a solar-energy future, at the present time hardly possible for political reasons alone, must not let us lose sight of what solar energy can presently contribute to our energy requirements. Fortunately, there is enormous potential in the industrialized countries themselves, because, otherwise, the technologies required for use in the deserts would hardly be developed. As an example of an industrialized country, we assess the situation for solar generation of electrical power in Germany.

In Germany, there are about 80 million people living in an area of 357 000 km^2. This gives a population density of 226 persons/km^2, with 4425 m^2 available to each person. In Germany, the Sun supplies about 1000 kWh/(a m^2), that is, 115 W m^{-2} averaged over the year. Over the area of 4425 m^2 per person, the Sun supplies around 500 kW. Compared with this, the current power requirement of 5.7 kW per person, 0.76 kW per person of this as electrical power, appears as almost negligibly small. Not all of this area will be used with high efficiency. The Germans allow themselves the 'luxury' of using 180 000 km^2, that is, half of the total area of Germany, to satisfy an energy requirement of merely 0.1 kW per person. This, however, is our most important energy requirement, our food, produced on agricultural farmland and meadowland.

To satisfy the electrical power requirements with solar cells, assuming an efficiency of 20%, which will become possible in the near future, we would need an area of 33 m^2 per person. This is very nearly as much as the average area of 35 m^2 living space available per person in Germany. Industrial buildings additionally account for at least the same area. Assuming three-story buildings on average gives a floor space of 23 m^2 per person for buildings in Germany. The roof surface areas will be somewhat greater. Only those oriented toward the north are unsuitable for solar cells. Furthermore, especially with multistoried buildings, the wall areas oriented toward the south are very well suited.

From this estimate, we can see that the areas at and on top of existing buildings are already sufficient for nearly covering our present energy requirements by the use of solar cells, even in a country that does not receive the most sunshine. There is no reason whatever to speak of replacing forests with solar cells. Since there is more sunshine in summer than in winter and more energy is needed in winter than in summer, it is necessary to store energy in the summer for use in winter. This problem remains to be solved, and will certainly entail storage losses. This, of course, makes additional roof area necessary. At present, though, a large amount of electrical power is wasted in producing low-temperature heat for heating buildings and for hot water. We could, in fact, manage with considerably less electrical power without any loss of comfort. Even for the relatively poor amount of sunshine in Germany, the enormous potential of solar energy for generating power without harm to the environment fully justifies the most intensive efforts to develop a solar-energy economy.

In view of the high population density and high rate of power consumption in industrialized countries like Germany, it seems more probable that not all of the power requirements will be met by utilizing only the solar energy captured in these countries. Even in an age of solar energy it is, in fact, more likely that industrialized countries will import energy from sunnier and less heavily populated countries, in very much the same way as today.

Solutions

Chapter 1

1.1 The greenhouse effect is caused in both cases by a layer that functions as a selective absorber being transparent for the incoming sunlight but nontransparent for the thermal radiation from the Earth's surface. In (a), this is caused by the plastic or glass on the roof and in (b), by certain gas molecules (three-atomic molecules like CO_2, N_2O and others) in the atmosphere.

1.2 Burning methane leads to a higher amount of CO_2 because the (mass) amount of carbon dioxide m_{CO_2} produced corresponds to 2.75 m_{CH_4} in the case of methane and to 1.47 m_{CH_2O} in the case of carbohydrates.

1.3 1 kg of coal equivalent corresponds to 8.2 kWh. On this amount of energy, the computer can be operated for 164 h or approximately 6.8 days.

1.4 For case (a) the function that describes the global human energy consumption is given by

$$\text{consumption}(t) = 15 \times 10^9\,(1 + t/35\text{a})\ \text{t coal eq./a}$$

In case (b) it is

$$\text{consumption}(t) = 15 \times 10^9 \exp\left[t \ln(2)/40\text{a}\right]\ \text{t coal eq./a}$$

In both cases,

$$\int_0^{t_{\text{end}}} \text{consumption}(t)\,\mathrm{d}t \stackrel{!}{=} 2 \times 10^{12}\ \text{t coal eq.}$$

has to be solved. This results in 272.4 years for case (a) and 183.7 years for case (b).

1.5 If all oxygen would have been produced by the reduction of CO_2 to methane, we have to consider the mass ratio $m_{CH_4} = \frac{16}{32} m_{O_2}$. The amount of methane would be $m_{CH_4} = 5 \times 10^{14}$ t.

Physics of Solar Cells: From Basic Principles to Advanced Concepts. Peter Würfel

ISBN: 978-3-527-40857-3

1.6 Venus is located at 0.723 AU (astronomical units) from the Sun, that is, 0.723 times the average distance $d_{\text{Earth-Sun}}$ between Earth and Sun. Therefore, the intensity reaching the outside of Venus is given by:

$$j_{\text{E, Venus}} = j_{\text{E, Earth}} \left(\frac{d_{\text{Earth-Sun}}}{d_{\text{Venus-Sun}}} \right)^2 = 2.49 \text{ kW m}^{-2}$$

Owing to the high reflectance of $r_v = 0.7$ the intensity reaching the surface is $j_{\text{E, Venus}} = 0.746 \text{ kW m}^{-2}$.

The temperature at the surface of Venus as a black body without greenhouse effect would be

$$T_v = \sqrt[4]{\frac{j_{\text{E, Venus}}}{4\sigma}} = 239.5 \text{ K} = -33.7\,^\circ\text{C}$$

In fact, the temperature at the surface of Venus is 475 °C. Accordingly, the number of fire screens is

$$n = \left(\frac{T_{\text{v, greenhouse}}}{T_v} \right)^4 - 1 = 94.2$$

Chapter 2

2.1 A black body is characterized by the fact that its absorptance (and hence also its emittance) is $a(\hbar\omega) \equiv 1$ for all photon energies $\hbar\omega$. It is best realized by a small opening in a large cavity.

2.2 According to Equation 2.27 the energy current density per solid angle from a black body is independent of the angle under which the black body is seen. The reduction of the received energy current caused by tilting the surface is thus compensated by an equal reduction of the apparent size of the surface area (i.e. the projection onto a plane perpendicular to the viewing direction). This phenomenon is observed as long as the emitting surface is resolved as an extended object. If the emitter is seen as a point source, only the variation of the energy current is registered but not the variation in apparent size of the emitting surface. For example, the different phases of Venus change in brightness when viewed with the naked eye, in contrast to observing it through a telescope, which resolves the bright and dark parts of the surface.

2.3 The number of states per volume for all photon energies $\leq \hbar\omega$ is given by Equation 2.5:

$$N_\gamma / V = \frac{(8\pi/3)\,(\hbar\omega)^3}{h^3 (c_0/n)^3}$$

For 1.5 eV $\leq \hbar\omega \leq$ 3 eV the number of states per volume is 10^{14} cm^{-3} for case (a) and 4.45×10^{15} cm^{-3} for case (b).

(c) From Equation 2.15 we know that the density of photons for *all* photon energies is

$$n_\gamma = \frac{(kT)^3}{4\pi^3\hbar^3(c_0/n)^3} \underbrace{\int_0^\infty \frac{x^2\,\mathrm{d}x}{\mathrm{e}^x - 1}}_{2.40411} 4\pi = \frac{2.40411\,k^3}{\pi^2\hbar^3(c_0/n)^3} T^3$$

Hence, in order to determine the density of photons for photon energies 1.5 eV $\leq \hbar\omega \leq$ 3 eV, we have to replace the limits of the integration by $x_1 = 1.5\ \mathrm{eV}/kT$ and $x_2 = 3\ \mathrm{eV}/kT$, respectively:

$$n_\gamma = \frac{(kT)^3}{4\pi^3\hbar^3(c_0/n)^3} \int_{x_1}^{x_2} \frac{x^2\,\mathrm{d}x}{\mathrm{e}^x - 1} 4\pi$$

As both the photon density n_γ and the number of photon states per volume N_γ/V depend on the refractive index n in the same way, the resulting temperature at which the total number of photons is larger than 10^{-4} of all photon states, does not depend on the particular material and is therefore the same for vacuum and for silicon.

For the photon energy range 1.5 eV $\leq \hbar\omega \leq$ 3 eV, this temperature is $T = 2582$ K.

2.4 The Boltzmann distribution differs from the Bose–Einstein distribution by less than 10^{-2} for photon energies larger than

(a) $\hbar\omega = 0.119$ eV at $T = 300$ K

(b) $\hbar\omega = 2.302$ eV at $T = 5800$ K.

In both cases, this threshold photon energy is $\hbar\omega = 4.6\,kT$. The Boltzmann distribution is thus a reasonable approximation of the Bose–Einstein distribution for $\hbar\omega \geq 4.6\,kT$.

2.5 The reflectance for normal incidence is given by

$$r(\alpha = 0) = \left(\frac{n_{\mathrm{air}}}{n_{\mathrm{Si}}}\right)^2 = 30.9\%$$

The component polarized perpendicular to the plane of incidence is reflected according to

$$r_\perp(\alpha) = \left(\frac{\sin(\alpha - \beta)}{\sin(\alpha + \beta)}\right)^2$$

and the component polarized parallel to the plane of incidence according to

$$r_{\parallel}(\alpha) = \left(\frac{\tan(\alpha - \beta)}{\tan(\alpha + \beta)}\right)^2$$

with $\beta = \arcsin\left(\frac{n_{\text{air}}}{n_{\text{Si}}} \sin\alpha\right)$. Hence the reflectance is 31.1% in case (b) and 46.7% in case (c).

2.6 **(a)** On a March 21, at noon, the Sun is incident normal to the surface at the equator. To receive an intensity of 750 W m^{-2}, the angle between the surface normal of the solar cell and the incoming radiation has to be $\alpha = \arccos(750\ \text{W m}^{-2}/1000\ \text{W m}^{-2}) = 41.4^\circ$. With the roof tilted by $\beta = 25^\circ$ toward the south, that is, toward the direction of the incoming light, the latitude is $\gamma = \alpha + \beta = 66.4^\circ$.

(b) As the Sun moves in a plane called the ecliptic, guiding of the module around a single axis perpendicular to the ecliptic assures that the module is always perpendicular to the incident radiation. For a March 21, the axis of the guiding system is parallel to the axis of the Earth and therefore $90^\circ - 66.4^\circ = 33.6^\circ$ tilted against the surface normal at 66.4° latitude.

Without a tracking system, the incident intensity at noon is 1000 W m^{-2} × cos(α) and varies during the day according to $\cos[\delta(t)]$, where $\delta(t)$ is the so-called hour-angle, which (on a March 21) varies in 12 h from $-\pi/2$ at Sunrise to $\pi/2$ at sunset. The energy output W of the module is

$$W = 1\ \text{kW m}^{-2} \times 0.15 \cos(\alpha)\ 12\ \text{h}/\pi \int_{-\pi/2}^{\pi/2} \cos(\delta)\, d\delta = 0.86\ \text{kWh m}^{-2}$$

With the tracking system, the energy output for the whole day is simply given by 1000 W m^{-2} × 12h × 0.15 = 1.8 kWh m^{-2}.

(To obtain an optimal output throughout the whole year, the guiding axis has to be adjusted according to the change of the inclination of the ecliptic with respect to the surface normal.)

2.7 For the mean photon energy of absorbed photons $\langle\hbar\omega_{\text{abs}}\rangle$, we have to divide the absorbed energy current density $j_{E,\text{abs}}$ in Equation 2.39 by the absorbed photon current density $j_{\gamma,\text{abs}}$ in Equation 2.40. $\varepsilon_G \gg kT$ allows to approximate these expressions by considering only the first element in the sum ($i = 1$).

$$\langle\hbar\omega_{\text{abs}}\rangle = kT\,\frac{x_G^3 + 3x_G^2 + 6x_G + 6}{x_G^2 + 2x_G + 2} = kT\left(x_G + 1 + 2/x_G + O\left(x_G^{-3}\right)\right)$$

where $x_G = \varepsilon_G/kT$. Thus, $\langle\hbar\omega_{\text{abs}}\rangle \approx \varepsilon_G + kT$.

The condition $\varepsilon_G \gg kT$ is very well fulfilled for a typical semiconductor like Si or GaAs in equilibrium with the 300 K radiation from the environment.

The reader may wonder how the absorbed photons with a mean energy of $\varepsilon_G + kT$ can be in equilibrium with electron–hole pairs with a mean energy of $\varepsilon_G + 3kT$, as derived in Equations 3.20 and 3.24.

Chapter 3

3.1 Doping with donors increases the concentration of electrons in the conduction band. A donor atom has a valency that is larger than the valency of the lattice atoms. It is important that the energy level (of one) of the electron(s) not needed for the chemical bond is close to the conduction band. An analogous argument applies for acceptors and the concentration of holes. Doping increases the conductivity, however, it reduces the mobility by introducing coulomb scattering by charged centers (as deviation from the periodic potential of the lattice ions).

3.2 Since the generation rate of electron–hole pairs by the 300 K radiation from the environment is not changed by doping, the equilibrium recombination rate cannot change either. As this radiative recombination rate is proportional to the product $n_e n_h$, an increase in n_e leads to a decrease in n_h.

3.3 The condition is that the thermalization time is much shorter than the lifetime, such that both the electrons in the conduction band and the holes in the valence band form ideal gases with a defined temperature.

3.4 Consider an impurity with a capture cross section much larger for electrons than for holes, $\sigma_e \gg \sigma_h$. In steady state, from Equation 3.66, we get

$$G_{e,\text{imp}} = R_{e,\text{imp}} - R_{h,\text{imp}} + G_{h,\text{imp}}$$

Because $\sigma_e \gg \sigma_h$, both exchange rates with the conduction band are much larger than the exchange rates with the valence band: $G_{e,\text{imp}}$, $R_{e,\text{imp}} \gg R_{h,\text{imp}}$, $G_{h,\text{imp}}$.

In this case, the occupation of the impurity is determined by the quasi-Fermi distribution for the conduction band (ε_{FC}), although the impurity level may lie in between ε_{FC} and ε_{FV}. For this reason, demarcation levels are introduced, which define the energy range in which the occupation of impurity levels is determined by the kinetics.

3.5 The electrical conductivity as resulting from Equations 5.3 and 5.4 is

$$\sigma = \sigma_e + \sigma_h = e\, n_e\, b_e + e\, n_h\, b_h$$

In an intrinsic semiconductor, the increase of the densities of free charge carriers, that is, electrons in the conduction band and holes in the valence band, rises exponentially with increasing temperature, according to Equations 3.12 and 3.15 (the dependence of the effective density of states on temperature is a minor correction). In a metal, there are only electrons and their concentration does not depend on temperature. On the other hand, the mobility is reduced with increasing temperature due to phonon scattering of the charge carriers. In a semiconductor ($\langle \varepsilon_{kin} \rangle = 3/2\, kT$), the mean thermal velocity $\langle v_{th} \rangle$ rises with $\sqrt{T}$. The concentration of phonons n_Γ, by which the charge carriers are scattered depends on temperature according to

$$n_\Gamma = N_\Gamma \left(\exp[\hbar\Omega/kT] - 1\right)^{-1} \overset{kT \gg \hbar\Omega}{\approx} N_\Gamma \left(1 + \hbar\Omega/kT - 1\right)^{-1} \approx \frac{N_\Gamma\, kT}{\hbar\Omega}$$

N_Γ is the density of states of the phonons and $\hbar\Omega$ is their energy.

The collision time τ_c is proportional to $1/\left(n_\Gamma \langle v_{th} \rangle\right)$ and thus the mobility in a semiconductor varies as $b_{e,h} \sim T^{-3/2}$.

If a semiconductor is doped, the temperature dependence of its conductivity exhibits three distinct temperature regimes. Figure 3.12 shows how the charge carrier densities depend on temperature. For temperatures relevant for solar cells (around room temperature), the concentration of the majority carriers is independent of temperature. The conductivity thus reflects the temperature dependence of their mobility.

The overall temperature dependence of the conductivity of an intrinsic semiconductor is, however, dominated by the exponential dependence of the charge carrier densities.

In a metal, $\langle v_{th} \rangle$ is independent of temperature, thus $b \sim T^{-1}$. Since the electron concentration is independent of temperature, the conductivity of a metal is proportional to T^{-1} as well.

3.6 An unoccupied conduction band would definitely minimize the energy E, but the quantity to minimize is the free energy $F = E - TS$. Since the entropy S is zero for an unoccupied conduction band and rises, for small concentrations, more strongly with increasing electron concentration than the energy, the minimum of F is found for a partially occupied conduction band. This topic is analyzed quantitatively in Problem 3.10.

3.7 As was derived in Equation 3.7 the density of states for a three-dimensional electron gas is

$$D_{e,3-\mathrm{dim}}(\varepsilon_e) = 4\pi \left(\frac{2\, m_e^*}{h^2}\right)^{3/2} (\varepsilon_e - \varepsilon_C)^{1/2}$$

It rises with the square root of the (kinetic) energy of the electrons.

In the one-dimensional case for a chain of length L, $\Delta p = h/L$ and the number of states with momentum $|p'| \leq |p|$ is $N(|p|) = 4|p|L/h$. This leads to

$$D_{\mathrm{e,1-dim}}(\varepsilon_{\mathrm{e}}) = 2\left(\frac{2\,m_{\mathrm{e}}^{*}}{h^{2}}\right)^{1/2}(\varepsilon_{\mathrm{e}} - \varepsilon_{\mathrm{C}})^{-1/2}$$

In the two-dimensional case for an area of L^2, we find $N(|p|) = 2\pi p^2 L^2/h^2$. The density of states is

$$D_{\mathrm{e,2-dim}}(\varepsilon_{\mathrm{e}}) = 2\pi\left(\frac{2\,m_{\mathrm{e}}^{*}}{h^{2}}\right)$$

3.8 The absorption coefficient $\alpha(\hbar\omega) \sim (\hbar\omega - \varepsilon_{\mathrm{G}})^{1/2}$ if the combined density of states varies as $D_{\mathrm{comb}}(\hbar\omega) \sim \sqrt{\varepsilon_{\mathrm{kin}}}$ and the transitions are direct.

3.9 The Boltzmann distribution differs from the Fermi–Dirac distribution by no more than 10^{-2} if the quasi-Fermi energy is at least $4.6\,kT$ lower than the conduction band edge for all temperatures.

Compare with Problem 2.4.

3.10 (a) We know that electrons and holes in a nondegenerate semiconductor form ideal gases with mean energies of $\langle\varepsilon_{\mathrm{e}}\rangle = \varepsilon_{\mathrm{C}} + 3/2\,kT$ and $\langle\varepsilon_{\mathrm{h}}\rangle = -\varepsilon_{\mathrm{V}} + 3/2\,kT$. For this reason, we can say that, starting from $F^* = 0$ (empty conduction band), the energy E per volume rises by $\varepsilon_{\mathrm{C}} - \varepsilon_{\mathrm{V}} + 3kT = \varepsilon_{\mathrm{G}} + 3kT$ for every electron "brought" into the conduction band, leaving behind a hole in the valence band. The entropy per volume is calculated from the Sackur–Tetrode equation (3.35), so the free energy per volume is

$$F^*(n_{\mathrm{e,h}}) = \frac{3\,kT + \varepsilon_{\mathrm{G}}}{2}(n_{\mathrm{e}} + n_{\mathrm{h}}) - kT\left\{n_{\mathrm{e}}\left[\frac{5}{2} + \ln\left(\frac{N_{\mathrm{C}}}{n_{\mathrm{e}}}\right)\right] + n_{\mathrm{h}}\left[\frac{5}{2} + \ln\left(\frac{N_{\mathrm{V}}}{n_{\mathrm{h}}}\right)\right]\right\}$$

For an intrinsic semiconductor, $n_{\mathrm{e}} = n_{\mathrm{h}} = n_{\mathrm{i}}$, which leads to

$$\begin{aligned} F^*(n_{\mathrm{i}}) &= n_{\mathrm{i}}\left\{\varepsilon_{\mathrm{G}} + 3kT - kT\left[5 + \ln\left(\frac{N_{\mathrm{C}}}{n_{\mathrm{i}}}\right) + \ln\left(\frac{N_{\mathrm{V}}}{n_{\mathrm{i}}}\right)\right]\right\} \\ &= n_{\mathrm{i}}\left\{\varepsilon_{\mathrm{G}} - kT\left[2 + \ln\left(\frac{N_{\mathrm{C}}}{n_{\mathrm{i}}}\right) + \ln\left(\frac{N_{\mathrm{V}}}{n_{\mathrm{i}}}\right)\right]\right\} \\ &= n_{\mathrm{i}}\left\{\varepsilon_{\mathrm{G}} - kT\left[2 + \ln\left(\frac{N_{\mathrm{C}}N_{\mathrm{V}}}{n_{\mathrm{i}}^{2}}\right)\right]\right\} \end{aligned}$$

To find the minimum of F^* as a function of the electron (hole) concentration, we set the derivative to zero

$$0 = \frac{\partial F^*(n_i)}{\partial n_i} = \varepsilon_G - kT\left[2 + \ln\left(\frac{N_C N_V}{n_i^2}\right)\right] - n_i\, kT\left(-\frac{2}{n_i}\right)$$

$$= \varepsilon_G - kT\ln\left(\frac{N_C N_V}{n_i^2}\right) \Longrightarrow n_i^2 = N_C\, N_V \exp\left(-\frac{\varepsilon_G}{kT}\right)$$

As expected, this result is exactly the same as the one for n_i^2 in Equation 3.17 and reproduces, therefore, the result obtained by Fermi–Dirac (or Boltzmann) statistics.
Inserting the result for n_i obtained above into the equation for F^* leads to a very simple expression (refer to Equation 3.17):

$$F^*(n_i) = n_i\left\{\varepsilon_G - kT\left[2 + \ln\left(\frac{N_C\, N_V}{n_i^2}\right)\right]\right\} = -2n_i\, kT$$

For the given parameters the free energy per volume reaches its minimum $F^* = -3.52 \times 10^8$ eV cm^{-3} for an electron (hole) concentration of $n_{e,h} = 6.798 \times 10^9$ cm^{-3}.

(b) Here, we have to regard the electrons that are transferred from states with an electron energy of $\varepsilon_{e,1} = \varepsilon_V - \varepsilon_0$ to states with an energy of $\varepsilon_{e,2} = \varepsilon_C + \varepsilon_0$. We then have to integrate over all these states, i.e. using the approximation made in Equation 3.12 from $\varepsilon_0 = 0$ to $\varepsilon_0 = \infty$.

This is equivalent to the situation with many narrow-band semiconductors with different values of ε_G, but all with the same Fermi energy ε_F. For every semiconductor, we have to find the minimum for F^* as a function of n_e and n_h and sum over all these contributions to get the total value for F^*.

From statistics, we know that the number of possibilities Ω to distribute n undistinguishable electrons (or holes) in N states is given by

$$\Omega = \binom{n}{N} = \frac{N!}{n!(N-n)!}$$
$$\approx N\ln(N) - n\ln(n) - (N-n)\ln(N-n)$$

The so-called Stirling approximation is a good approximation for large numbers N, n.

For the following calculation, we divide the bands into intervals $\Delta\varepsilon$ and determine the number of states per volume ΔN_l for intervals centered at an energy

$$\varepsilon_{e,l} = \varepsilon_C + (l + 1/2)\,\Delta\varepsilon$$

and

$$\varepsilon_{h,l} = -\varepsilon_V + (l + 1/2)\,\Delta\varepsilon$$

respectively, with

$$\Delta N_{\mathrm{e,h},l}(\varepsilon_{\mathrm{e,h},l}) = D_{\mathrm{e,h}}(\varepsilon_{\mathrm{e,h},l})\Delta\varepsilon$$

$N_C = 3 \times 10^{19}\ \mathrm{cm}^{-3}$ and $N_V = 1 \times 10^{19}\ \mathrm{cm}^{-3}$ correspond to effective masses of $m_e^* = 1.1264\, m_e$ and $m_h^* = 0.5415\, m_e$, respectively. These determine $D_{\mathrm{e,h}}(\varepsilon_{\mathrm{e,h},l})$.

We obtain the free energy per volume for every $\Delta\varepsilon$ centered at ε_l

$$\begin{aligned}
\Delta F_l^* = \Delta E_l - T(\Delta S_{\mathrm{e},l} + \Delta S_{\mathrm{h},l}) &= \Delta n_l[\varepsilon_G + 2(l + 1/2)\Delta\varepsilon] \\
&- kT[\Delta N_{l,\mathrm{e}} \ln(\Delta N_{\mathrm{e},l}) - \Delta n_l \ln(\Delta n_l) \\
&- (\Delta N_{\mathrm{e},l} - \Delta n_l)\ln(\Delta N_{\mathrm{e},l} - \Delta n_l)] \\
&- kT[\Delta N_{l,\mathrm{h}} \ln(\Delta N_{\mathrm{h},l}) - \Delta n_l \ln(\Delta n_l) \\
&- (\Delta N_{\mathrm{h},l} - \Delta n_l)\ln(\Delta N_{\mathrm{h},l} - \Delta n_l)]
\end{aligned}$$

We again made use of $\Delta n_{\mathrm{e},l} = \Delta n_{\mathrm{h},l} = \Delta n_l$, being valid for an intrinsic semiconductor also in the case of $m_e^* \neq m_h^*$.

After having determined the minimum of $\Delta F_l^*(\Delta n_l)$, we have to sum over all l to get the correct results for F^* and for $n_e = n_h = n_i$.

$$F^* = \sum_{l=0}^{\infty} \Delta F_l^* \quad \text{and} \quad n_e = \sum_{l=0}^{\infty} \Delta n_l$$

Again, we can determine the minimum of F^* analytically:

$$\begin{aligned}
0 = \frac{\partial \Delta F^*(\Delta n_l)}{\partial \Delta n_l} &= \varepsilon_C - \varepsilon_V + (2l + 1)\Delta\varepsilon \\
&- kT\left[-\ln(\Delta n_l) - 1 + \ln\left(\Delta N_{\mathrm{e},l} - \Delta n_l\right) + 1\right. \\
&\left.- \ln(\Delta n_l) - 1 + \ln\left(\Delta N_{\mathrm{h},l} - \Delta n_l\right) + 1\right] = \varepsilon_G + (2l + 1)\Delta\varepsilon \\
&- kT\left[\ln\left(\frac{\Delta N_{\mathrm{e},l} - \Delta n_l}{\Delta n_l}\right) + \ln\left(\frac{\Delta N_{\mathrm{h},l} - \Delta n_l}{\Delta n_l}\right)\right]
\end{aligned}$$

- For $\Delta N_{\mathrm{e},l} = \Delta N_{\mathrm{h},l} = \Delta N_l$ ($m_e^* = m_h^*$) follows

$$\begin{aligned}
0 &= \frac{\partial \Delta F^*(\Delta n_l)}{\partial \Delta n_l} \\
&= \varepsilon_G + (2l + 1)\Delta\varepsilon - 2kT\left[\ln\left(\frac{\Delta N_l - \Delta n_l}{\Delta n_l}\right)\right]
\end{aligned}$$

and finally

$$\Delta n_l = \frac{\Delta N_l}{\exp\left[\left(\varepsilon_G + (2l+1)\Delta\varepsilon\right)/2kT\right] + 1}$$
$$\stackrel{!}{=} \frac{\Delta N_l}{\exp\left[\left(\varepsilon_C + (l+1/2)\Delta\varepsilon - \varepsilon_F\right)/kT\right] + 1}$$

from which follows that $\varepsilon_F = (\varepsilon_C + \varepsilon_V)/2$, as expected. This is exactly the Fermi–Dirac formula for the electron concentration Δn_l at the energy $\varepsilon_{e,l} = \varepsilon_C + (l + 1/2)\Delta\varepsilon$.

- For $\Delta N_{e,l} \neq \Delta N_{h,l}$ $(m_e^* \neq m_h^*)$

$$\Delta n_l = \Delta N_{e,l} \left(\exp\left[\left(\varepsilon_{e,l} - \varepsilon_F\right)/kT\right] + 1\right)^{-1}$$

and

$$\Delta n_l = \Delta N_{h,l} \left(\exp\left[\left(\varepsilon_F + \varepsilon_{h,l}\right)/kT\right] + 1\right)^{-1}$$

it follows that

$$\varepsilon_F = \frac{1}{2}(\varepsilon_{e,l} - \varepsilon_{h,l}) + \frac{1}{2}kT \ln\left(\frac{\Delta N_{h,l} - \Delta n_l}{\Delta N_{e,l} - \Delta n_l}\right)$$

With the Boltzmann approximation, i.e. $\Delta N_{e,h,l} - \Delta n_l \approx \Delta N_{e,h,l}$ we get the result of Equation 3.18

$$\varepsilon_F = \frac{1}{2}(\varepsilon_{e,l} - \varepsilon_{h,l}) + \frac{1}{2}kT \ln\frac{\Delta N_{h,l}}{\Delta N_{e,l}}$$

Using the same approximation for the conditional equation for the minimum of F^*, we get

$$0 = \frac{\partial \Delta F^*(\Delta n_l)}{\partial \Delta n_l} = \varepsilon_G + (2l+1)\Delta\varepsilon$$
$$- kT\left[\ln\left(\frac{\Delta N_{e,l} - \Delta n_l}{\Delta n_l}\right) + \ln\left(\frac{\Delta N_{h,l} - \Delta n_l}{\Delta n_l}\right)\right]$$
$$\approx \varepsilon_G + (2l+1)\Delta\varepsilon - kT\left[\ln\left(\frac{\Delta N_{e,l}\,\Delta N_{h,l}}{\Delta n_l^2}\right)\right]$$

Again, we see that Δn_l equals the intrinsic carrier concentration $\Delta n_{i,l}$

$$\Delta n_l = \sqrt{\Delta N_{e,l}\,\Delta N_{h,l}}\,\exp\left[-\frac{\varepsilon_G + (2l+1)\Delta\varepsilon}{2kT}\right] = \Delta n_{i,l}$$

With the above equation for the Fermi energy in Boltzmann approximation, this can be rewritten as

$$\Delta n_l = \Delta N_{e,l}\left(\exp\left[\frac{\begin{array}{c}\varepsilon_C+(l+1/2)\Delta\varepsilon\\ -\varepsilon_C/2-\varepsilon_V/2\end{array}}{kT}\right]\sqrt{\Delta N_{e,l}/\Delta N_{h,l}}\right)^{-1}$$

$$= \Delta N_{e,l}\left(\exp\left[\frac{\begin{array}{c}\varepsilon_C+(l+1/2)\Delta\varepsilon-\varepsilon_C/2-\varepsilon_V/2\\ -kT/2\,\ln\left(\Delta N_{h,l}/\Delta N_{e,l}\right)\end{array}}{kT}\right]\right)^{-1}$$

$$= \Delta N_{e,l}\,\exp\left[-\left(\varepsilon_{e,l}-\varepsilon_F\right)/kT\right]$$

Without the Boltzmann approximation, the direct determination of the Fermi energy ε_F from the minimum of the free energy is a little complicated. Instead, we assume that we know the Fermi distribution for the holes, which gives

$$\varepsilon_{h,l}+\varepsilon_F = kT\,\ln\frac{\Delta N_{h,l}-\Delta n_l}{\Delta n_l}$$

we insert this into

$$0=\frac{\partial\Delta F^*(\Delta n_l)}{\partial\Delta n_l}$$

$$=\varepsilon_{e,l}+\varepsilon_{h,l}-kT\left[\ln\left(\frac{\Delta N_{e,l}-\Delta n_l}{\Delta n_l}\right)+\ln\left(\frac{\Delta N_{h,l}-\Delta n_l}{\Delta n_l}\right)\right]$$

$$=\varepsilon_{e,l}-\varepsilon_F-kT\ln\left(\frac{\Delta N_{e,l}-\Delta n_l}{\Delta n_l}\right)$$

and find the exact Fermi distribution for the electrons

$$\Delta n_l=\frac{\Delta N_{e,l}}{\exp\left[\left(\varepsilon_C+(l+1/2)\Delta\varepsilon-\varepsilon_F\right)/kT\right]+1}$$

Figure S.1a shows the free energy per volume F^* as a function of the electron (hole) concentration of intrinsic silicon. The entropy is given by Sackur–Tetrode. The minimum of F^* determines the state of chemical equilibrium between the electrons and the holes.

Figure S.1b shows the sum over the contributions from each energy interval up to an energy $\varepsilon_{e,l}$ to the total free energy F^* and to the total electron (hole) concentration in the state of chemical equilibrium defined by a minimum of F^*. One sees that for energies $\varepsilon_{e,l} >= \varepsilon_C + 0.25$ eV (and $\varepsilon_{h,l} >= -\varepsilon_V + 0.25$ eV) these contributions are negligibly small. The result in (b), involving the explicit calculation of the entropy, agrees well with the result in (a), using the entropy from the Sackur–Tetrode equation.

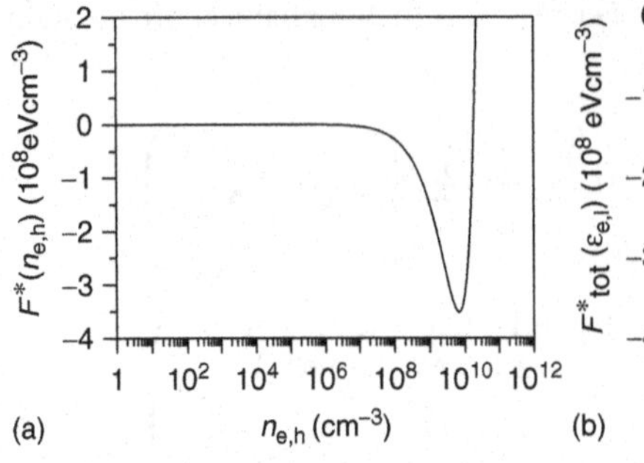

Figure S.1 (a) Free energy per volume F^* as a function of the electron (hole) concentration $n_{e,h}$ as a result of the Sackur–Tetrode equation (3.35). (b) F^* (solid line) and n_e (dotted line) as the sum of the contributions of all states with energies $\varepsilon_e \leq \varepsilon_{e,l}$ as a function of $\varepsilon_{e,l}$. In graph (a), effective densities of states $N_C = 3 \times 10^{19}$ cm^{-3} and $N_V = 1 \times 10^{19}$ cm^{-3} were assumed and in graph (b) the corresponding effective masses $m_e^* = 1.1264\, m_e$ and $m_h^* = 0.5415\, m_e$. For both, the temperature is $T = 300$ K.

3.11 The binding energy ε_{bind} for the ground and the excited states of an electron in a donor atom using the hydrogen approximation is

$$\varepsilon_{bind}(n) = \frac{m_e^* e^4}{2\,(4\pi\epsilon_0\epsilon)^2\,\hbar^2}\frac{1}{n^2}$$

with n being the quantum number for the state, i.e. $n = 1$ for the ground state and $n = 2, 3...$ for the excited states. For ϵ (Si) $= 11.9$ we get

(a) $\varepsilon_{bind}(1) = 0.104$ eV, $\varepsilon_{bind}(2) = 0.026$ eV, $\varepsilon_{bind}(3) = 0.012$ eV.

(b) $\varepsilon_{bind}(1) = 1.9 \times 10^{-3}$ eV, $\varepsilon_{bind}(2) = 4.8 \times 10^{-4}$ eV, $\varepsilon_{bind}(3) = 2.1 \times 10^{-4}$ eV.

(c) The Bohr radius r_B for the ground state ($n = 1$) is given by

$$r_B(1) = \epsilon_0\epsilon_{Si}\frac{h^2}{\pi m_e^* e^2}$$

For case (a), it is $r_B(1) \approx 0.6$ nm and for case (b), $r_B(1) \approx 31.5$ nm. For a mass density of crystalline Si of $\rho_{Si} = 2.3$ g cm^{-3} and a molar mass of $M_{Si} = 28.1$ g mol^{-1}, there are about 4.95×10^{22} atoms in 1 cm^{-3}. It follows that in a spherical volume with the Bohr radius of case (a) calculated above, there are about 32 atoms over which the electron is then smeared out. In case (b), the number of atoms is 5.1×10^6(!).

3.12 **(a)** Full ionization of the donors means $n_e = n_D$ and therefore $\varepsilon_C - \varepsilon_F = kT \ln \frac{N_C}{n_e} = 0.208$ eV.

(b) We have three equations for the three unknown quantities n_e, n_{D^+}, and ε_F. Two express the dependence of n_e and n_{D^+} on ε_F and a third one results from the fact that the semiconductor remains electrically

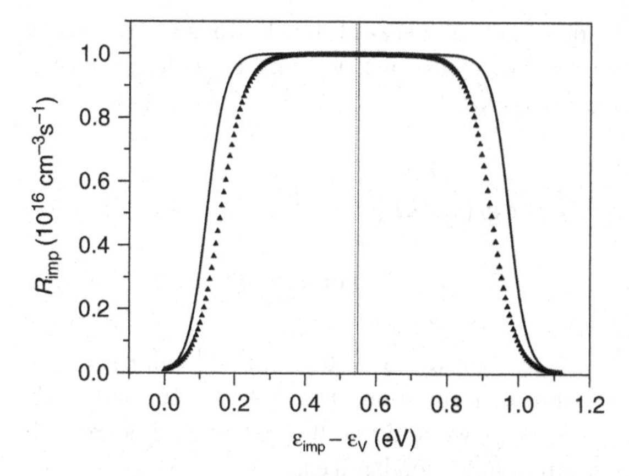

Figure S.2 Recombination rates R_{imp} from impurity states as a function of their energetic position for $T = 300$ K (solid line) and $T = 400$ K (triangles). The vertical solid line indicates the position of the intrinsic Fermi level with respect to the valence band maximum $\varepsilon_i - \varepsilon_V$ for $T = 300$ K, whereas the vertical dotted line represents $\varepsilon_i - \varepsilon_V$ for $T = 400$ K. The difference in ε_i for the two temperatures is rather small: $\varepsilon_i(300\text{ K}) - \varepsilon_V = 0.5458$ eV; $\varepsilon_i(400\text{ K}) - \varepsilon_V = 0.5411$ eV.

neutral upon doping ($e\left[n_D^+ + n_h - n_e\right] = 0$), see Equation 3.26 and following.

The results are $\varepsilon_C - \varepsilon_F = 0.218537$ eV using the Boltzmann approximation for n_e and $\varepsilon_C - \varepsilon_F = 0.218533$ eV with Fermi–Dirac.

(c) The temperature is $T(n_h \geq 0.5 n_e) \approx 639.8\,^\circ$C. At that temperature, 50.2% of the electrons in the conduction band come from ionized donor atoms.

3.13 The recombination rates are calculated using Equation 3.77. The recombination rates as a function of the impurity level for temperatures $T = 300$ K and $T = 400$ K are shown in Figure S.2.

Chapter 4

4.1 The efficiency η is given by $dj_\mu / dj_{E,\text{abs}}$, that is, the ratio of the extracted chemical energy current density $dj_\mu = dj_{eh}(\mu_e + \mu_h)$ and the absorbed energy current density $dj_{\gamma,\text{abs}}\,\hbar\omega$. Using Equation 4.10 we get

$$\eta = \left[dj_{\gamma,\text{abs}} - dj_\gamma^0 \exp\left(\frac{\mu_e + \mu_h}{kT}\right)\right](\mu_e + \mu_h) / \left(dj_{\gamma,\text{abs}}\,\hbar\omega\right)$$

As the first summand $dj_{\gamma,\text{abs}}$ does not depend on $(\mu_e + \mu_h)$, its contribution to the efficiency is linear in $(\mu_e + \mu_h)$. This linear dependence dominates the behavior of η as long as the radiative recombination rate $dj_\gamma^0 \exp\left(\frac{\mu_e+\mu_h}{kT}\right)$ is small.

4.2 For the change of the quasi-Fermi energy of the electrons ε_{FC} caused by illumination under open-circuit conditions, we first have to determine the absorbed photon current density

$$j_{\gamma,\text{abs}} = \frac{\Omega}{4\pi^3\hbar^3c^2}\int_{\varepsilon_G}^{\infty}\frac{(\hbar\omega)^2}{\exp(\hbar\omega/kT_{bb})-1}\,d\hbar\omega$$

The mean generation rate is $\overline{G} = j_{\gamma,\text{abs}}/d$. For $d = 300\ \mu\text{m}$, we find $\overline{G} = 1.13\times10^{19}\ \text{cm}^{-3}\ \text{s}^{-1}$.

(a) From $\overline{G} = R_{rad} = Bn_hn_e$ with $B = 3\times10^{-15}\ \text{cm}^3\ \text{s}^{-1}$ and $n_h = n_h^0 + n_e$ we find an electron concentration of $n_e = 4.13\times10^{16}\ \text{cm}^{-3}$. By using $\varepsilon_C - \varepsilon_{FC} = kT_0\ln\frac{n_e}{N_C}$ we find that the quasi-Fermi energy ε_{FC} lies 0.170 eV below the conduction band edge.

(b) From $\overline{G} = R_{aug} = Cn_hn_e(n_h + n_e)$ with $C = 10^{-30}\ \text{cm}^6\ \text{s}^{-1}$ and $n_h = n_h^0 + n_e$ we find an electron concentraton of $n_e = 3.67\times10^{15}\ \text{cm}^{-3}$. Accordingly, the quasi Fermi-energy ε_{FC} lies 0.233 eV below the conduction band edge.

4.3 To calculate the emitted photon current density (into an effective solid angle of π) as a function of the chemical potential of the emitted photons (which equals the chemical potential of the electron–hole pairs), we have to make use of Equation 3.102. Since for the problem, this equation has to be expressed in terms of the wavelength λ, we need the relations in Equation 2.10 $\hbar\omega = hc/\lambda$ and $d\hbar\omega = -hc/\lambda^2\,d\lambda$. The transformed equation is

$$\frac{dj_\gamma}{d\lambda} = \frac{2c\,\Omega}{\lambda^4}\frac{1}{\exp\left[(hc/\lambda - \mu_{e,h})/kT\right]-1}$$

(a) For $\mu_{eh} = 0.5$ eV, the emitted photon current density per wavelength $dj_\gamma/d\lambda = 1.64\times10^{-12}\ \text{cm}^{-2}\ \text{nm}^{-1}$.

(b) For $\mu_{eh} = 1.8$ eV, $dj_\gamma/d\lambda = 1.14\times10^{10}\ \text{cm}^{-2}\ \text{nm}^{-1}$.

Chapter 5

5.1 We start from $j_{Q,i} = -\frac{\sigma_i}{z_ie}\,\text{grad}\,\eta_i$, which for pure diffusive transport becomes

$$j_{Q,i} = -\frac{\sigma_i}{z_ie}\,\text{grad}\,\mu_i = -\frac{\sigma_i}{z_ie}\,\text{grad}\,kT\ln(n_i/n_0) = -\frac{\sigma_i}{z_ie}kT\frac{1}{n_i}\text{grad}\,n_i$$

Using $\sigma_i = z_i^2en_ib_i$ from Section 5.1.1, we get

$$j_{Q,i} = -z_ib_ikT\,\text{grad}\,n_i$$

As we know, this result is equivalent to Fick's law of diffusion, hence

$$-z_i b_i kT \text{grad}\, n_i \equiv -D_i \,\text{grad}\, n_i$$

which finally leads to

$$D_i = \frac{b_i\, kT}{e}$$

5.2 **(a)** It is $\sigma_h = z_h^2 e n_h b_h = e \cdot 10^{16}\ \text{cm}^{-3}\ 300\ \text{cm}^2\ (\text{Vs})^{-1} = 0.48\ (\Omega\ \text{cm})^{-1}$.

(b) With $E = j_Q/\sigma_h = 0.312\ \text{V cm}^{-1}$, it follows that the electric potential difference between the two ends of the semiconductor is $\Delta\varphi = 6.24\ \text{mV}$.

(c) If the current is caused by diffusion, the gradient of the hole density has to be constant throughout the sample, due to $\text{div}\, j_Q = 0$. We have $D_h = b_h kT/e = 7.76\ \text{cm}^2\ \text{s}^{-1}$. $j_Q = -e D_h \,\text{grad}\, n_h$ leads to $\text{grad}\, n_h = 1.2 \times 10^{17}\ \text{cm}^{-4}$. The linear profile of the hole concentration is then given by

$$n_h(x) \approx (1.12 \times 10^{16} - 2.4 \times 10^{15}\, x/d)\ \text{cm}^{-3}$$

The gradient of the electrochemical potential is independent of x only in case (a), where it trivially results in $\text{grad}\, \eta_h = eE = 312\ \text{meV cm}^{-1}$. In case (b), we have to consider that the conductivity is a function of x and thus $\text{grad}\, \eta_h$ as well:

$$\text{grad}\, \eta_h = \text{grad}\, \mu_h = \frac{kT}{n_h}\frac{\partial n_h}{\partial x} \approx (278.5 + 3.8\, x/d)\ \text{meV cm}^{-1}$$

The difference in chemical potential between $x = 0$ and $x = d$ is

$$\Delta\eta = \Delta\mu = kT \ln \frac{n_h(0)}{n_h(d)} = 6.27\ \text{meV}$$

5.3 The mean drift velocity of the electrons is given by $\langle v_{e,\text{drift}} \rangle = b_e(T)E$ and the mean thermal velocity by $\langle v_{e,\text{th}} \rangle = \sqrt{3kT/m_e}$.

The temperature for which $\langle v_{e,\text{drift}} \rangle$ is $0.001\ \langle v_{e,\text{th}} \rangle$ is $T = 481$ K.

5.4 The mobility of a compensated semiconductor is smaller than of a pure one because the electrons and holes collide with the impurities. Thus, the mean time between two collisions, $\tau_{c,i}$ is reduced, and, with it, the mobility b_i, see Equation 5.2. This effect is most pronounced if the impurities are charged.

Chapter 6

6.1 An improvement in the lifetimes of electrons and holes by a factor of 4 leads to a doubling of their diffusion lengths. In addition, the recombination rates are lowered by a factor of 4. As a result, the reverse saturation current density j_S is reduced by a factor of 2, see Equation 6.28.

6.2 It is

$$w = \sqrt{\frac{2\,\epsilon\epsilon_0}{\mathrm{e}} \frac{n_A + n_D}{n_A\, n_D} (\varphi^n - \varphi^p - V_a)}$$

with V_a being the applied voltage. The diffusion voltage is calculated with Equation 6.10 and is $\varphi^n - \varphi^p = (kT/\mathrm{e}) \ln(n_D n_A/n_i^2) = 0.9\ V$ for the given doping concentrations. For an applied voltage of $V_a = -2.3\ V$, the width of the space charge region is $w = 0.65\ \mu\mathrm{m}$.

6.3 The space charge capacitance per unit area is given by $C_{sc}/A = C_{sc}^* = |\partial Q_{sc}^*/\partial V|$. We have

$$Q_{sc}^* = \mathrm{e} n_D w_n = \sqrt{2\mathrm{e}\,\epsilon\epsilon_0 \frac{n_A n_D}{n_A + n_D} (\varphi^n - \varphi^p - V_a)}$$

and hence

$$C_{sc}^* = \frac{1}{2}\sqrt{\frac{n_A n_D}{n_A + n_D} \frac{2\mathrm{e}\,\epsilon\epsilon_0}{\varphi^n - \varphi^p - V_a}}$$

The capacitance for 0 V is $C_{sc}^* \approx 30\ \mathrm{nF\ cm^{-2}}$ and for an applied voltage of -4 V, it is $C_{sc}^* \approx 13\ \mathrm{nF\ cm^{-2}}$.

If C_{sc}^{*-2} is plotted against V_a (the so-called Mott–Schottky plot), it is possible to determine $\varphi^n - \varphi^p$ and the smaller of the dopant concentrations if either $n_D \gg n_A$ or vice versa. It is

$$\frac{1}{C_{sc}^{*2}} = \frac{1}{2\mathrm{e}\epsilon\epsilon_0} \left(\frac{4(n_A + n_D)}{n_A n_D} (\varphi^n - \varphi^p) - \frac{4(n_A + n_D)}{n_A n_D} V_a \right)$$

From the intercept $C_{sc}^{*-2} = 0$, we get $V_a = (\varphi^n - \varphi^p)$. From the slope, n_A, (n_D) can be determined if $n_D \gg n_A$, $(n_A \gg n_D)$.

6.4 The separation of the quasi-Fermi energies $\varepsilon_{FC} - \varepsilon_{FV}$ as a function of the impurity level ε_{imp} can be calculated from Equations 3.67 and 3.77. For $n_{e,imp}$ in Equation 3.67 we have to use the expression given by Equation 3.76. We then have two equations for the two unknown quantities n_e and n_h. From the result we can easily derive ε_{FC} and ε_{FV}. The results for a) $G = 10^{18}\ \mathrm{cm^{-3}\ s^{-1}}$ and b) $G = 5 \times 10^{20}\ \mathrm{cm^{-3}\ s^{-1}}$ are shown in Figure S.3.

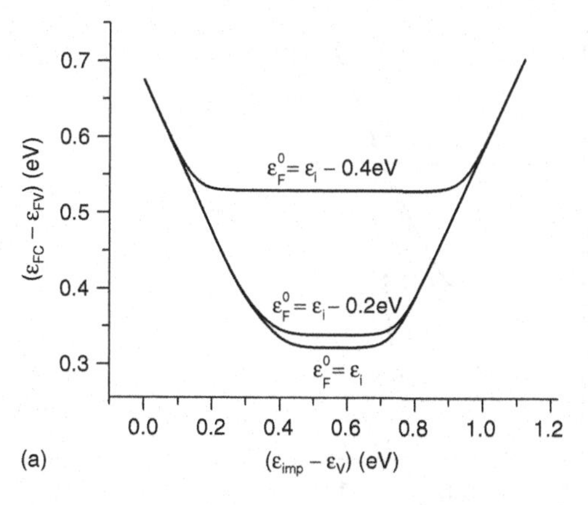

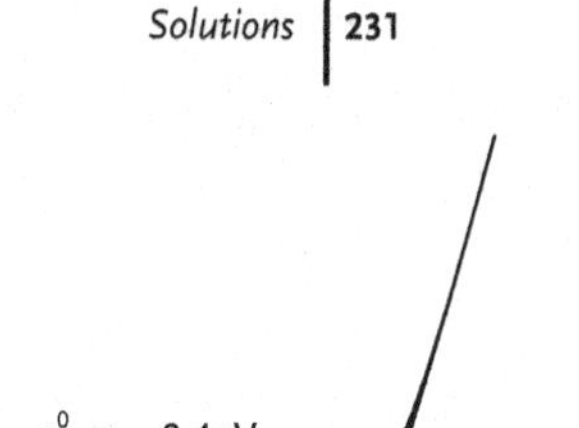

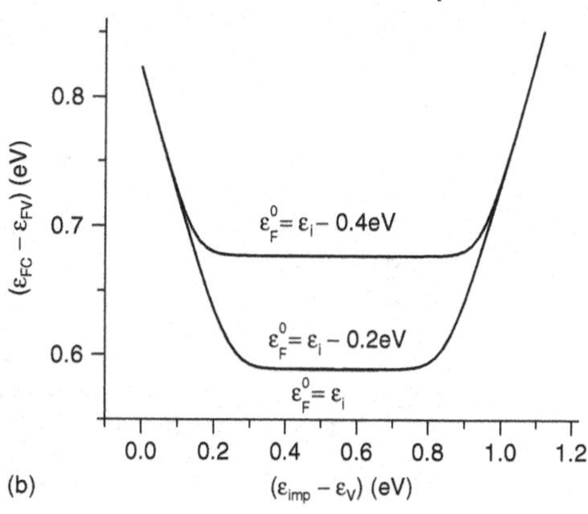

Figure S.3 Separation of the quasi-Fermi energies $\varepsilon_{FC} - \varepsilon_{FV}$ in Si as a function of the impurity level ε_{imp} for three different doping concentrations. (a) With a generation rate $G = 10^{18}$ cm^{-3} s^{-1}. (b) With $G = 5 \times 10^{20}$ cm^{-3} s^{-1}.

Chapter 7

7.1 Using Eq. (2.13) for the single diode model, we get

$$j_Q = j_S\left[\exp\left(\frac{e\left(V - j_Q\tilde{R}_s\right)}{kT}\right) - 1\right] + j_{sc} + \frac{V - j_Q\tilde{R}_s}{\tilde{R}_p}$$

This transcendent equation is solved numerically and the results are shown in Figure S.4.

Chapter 8

8.1 A solar cell produces electrical energy from the absorbed solar energy. Ideally, a conversion system must be able to absorb the entire solar spectrum. Absorption losses are avoided in the thermophotovoltaic arrangement in Section 8.3, or in the hot carrier cell in Section 8.4.2 or in a tandem configuration of many cells with different band gaps in Section 8.1. The first step of the conversion process is the production of chemical energy of electron–hole pairs. This chemical energy is given by the difference of the Fermi energies $\varepsilon_{FC} - \varepsilon_{FV}$ of two different Fermi distributions describing separately the occupation of the valence and of the conduction band. For these distributions, the temperature must be defined. The first step in the conversion process is, therefore, the thermalization of the electrons and holes. If this thermalization occurs over a broad energy range, energy is lost. This loss is strongly reduced by restricting the thermalization to

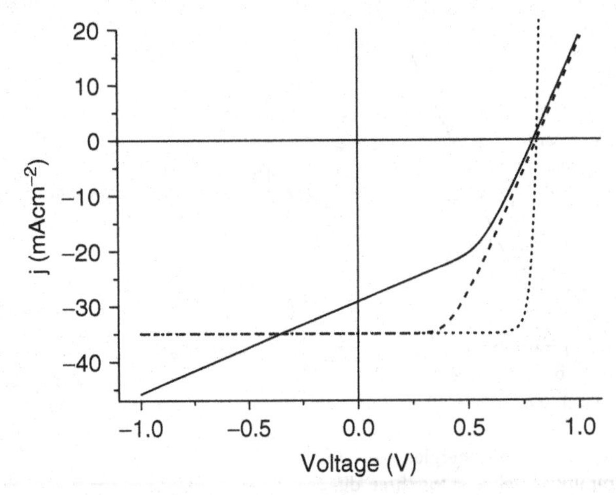

Figure S.4 Current–voltage characteristics of an ideal pn-junction with $\tilde{R}_s = 0$, $\tilde{R}_p = \infty$ (dotted line), one with $\tilde{R}_s = 10\,\Omega\text{cm}^2$, $\tilde{R}_p = \infty$ (broken line) and one with $\tilde{R}_s = 10\,\Omega\text{cm}^2$, $\tilde{R}_p = 50\,\Omega\text{cm}^2$ (solid line).

a narrow energy range, as by restricting already the generation of electron–hole pairs to a narrow energy range by a tandem configuration and by the filter in the thermophotovoltaic arrangement, or by thermalization in the energy-selective contacts of a hot carrier cell. The final conversion of chemical energy into electrical energy requires semipermeable membranes and a small resistance for the transport of the electrons and holes from where they are generated to the outside circuit. This ensures that the gradients of the quasi-Fermi energies, the driving forces for the transport, are small and little chemical energy is lost in the conversion to electrical energy.

8.2 **(a)** To calculate the different recombination rates from the given quantity n_h the electron concentration n_e has to be determined. It follows from the condition of charge neutrality for the semiconductor. For radiative and/or Auger recombination, this condition leads to the equation $n_e = n_D + n_h$ according to Equation 3.26. If impurities are involved, the equation reads $n_e = n_D + n_h - n_{e,imp}$ according to Equation 3.67. The results are shown in Figure S.5.

It can be seen from the graph that the impurity recombination rate determines the total rate for small values of n_h, corresponding to low illumination intensities. For higher values of n_h, the impurity recombination saturates, whereas the radiative and the Auger recombination become more important. Finally, for values $n_h \geq 3 \times 10^{16}\ \text{cm}^{-3}$, Auger recombination dominates the total recombination rate. Note that for the given parameters, radiative recombination is not dominant at any intensity.

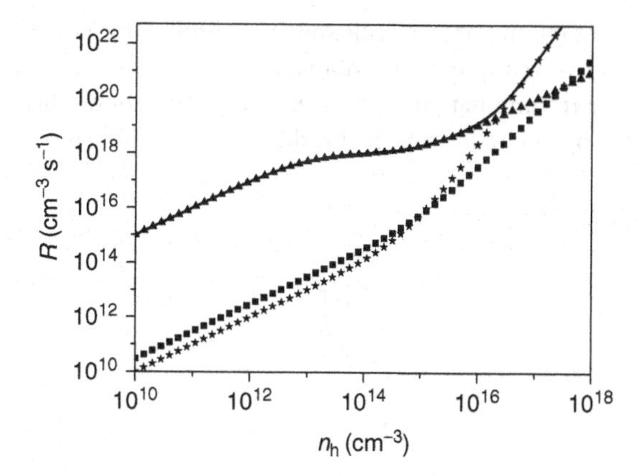

Figure S.5 The total recombination rate (solid line) as a function of the minority carrier concentration n_h and the different contributions from radiative (■), Auger (★), and impurity recombination (▲).

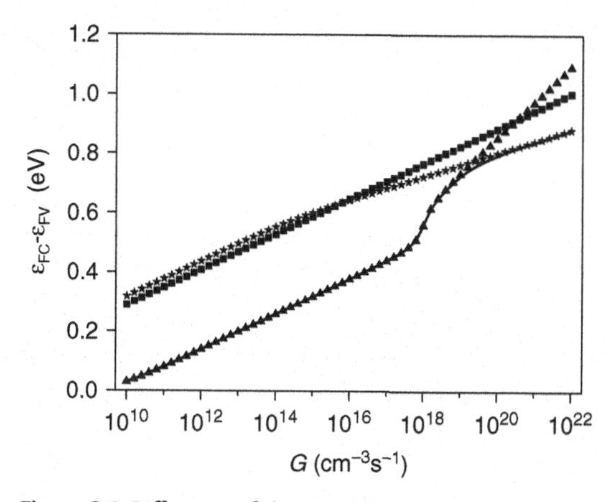

Figure S.6 Difference of the quasi-Fermi energies $\varepsilon_{FC} - \varepsilon_{FV}$ under open-circuit conditions as a function of the generation rate G. The recombination mechanism taken into account is either radiative (■), Auger (★), or impurity recombination (▲) or a combination of all three types of recombination (solid line).

(b) We can calculate the electron concentration n_e as a function of n_h and determine the quasi-Fermi energies and the recombination rates as a function of n_e and n_h. The other way is to determine n_e and n_h as a function of the recombination rates (which equal the generation rates) and determine ε_{FC} and ε_{FV} from n_e and n_h. The equation for charge neutrality is the same as in (a). The difference of the Fermi energies as a function of the generation rate for the different recombination mechanisms is shown in Figure S.6.

As is expected from (a), the open-circuit voltage is limited for small generation rates solely by impurity recombination, whereas for higher values of G, Auger recombination dominates. The graph shows the limitation of the open-circuit voltage by different recombination mechanisms.

Appendix

Fundamental Constants

Boltzmann constant	$k = 1.3807 \times 10^{-23}$ Ws K^{-1} = 8.617×10^{-5} eV K^{-1}
Planck constant	$h = 6.626 \times 10^{-34}$ Ws^2 = 4.136×10^{-15} eV s
	$\hbar = 1.0546 \times 10^{-34}$ Ws^2 = 6.582×10^{-16} eV s
Velocity of light	$c_{vac} = 2.998 \times 10^8$ m s^{-1}
Elementary charge	$e = 1.602 \times 10^{-19}$ As
Electron mass	$m_e = 9.109 \times 10^{-31}$ kg
Permittivity of free space	$\epsilon_0 = 8.85 \times 10^{-12}$ As V^{-1} m^{-1}
Stefan–Boltzmann constant	$\sigma = 5.67 \times 10^{-8}$ W m^{-2} K^{-4}
Solid angle of solar disc	$\Omega_s = 6.8 \times 10^{-5}$
	$\hbar\omega\lambda = hc_{vac} = 1.240$ eV μm
	$\frac{1}{4\pi^3\hbar^3 c_{vac}^2} = 5.04 \times 10^7 \frac{\text{W}}{(\text{eV})^4\ \text{m}^2}$

Units of Energy

$1\ \text{eV} = 1.602 \times 10^{-19}$ J
$1\ \text{J} = 1\ \text{Ws} = 1\ \text{N m}$
$1\ \text{kWh} = 3.6 \times 10^6$ J

Physics of Solar Cells: From Basic Principles to Advanced Concepts. Peter Würfel

ISBN: 978-3-527-40857-3

Material Constants at 300 K

	Ge	Si	GaAs
ε_G (eV)	0.66	1.12	1.42
χ (eV)	4.13	4.01	4.07
ϵ	16	11.9	13.1
N_C (cm^{-3})	1×10^{19}	3×10^{19}	5×10^{17}
N_V (cm^{-3})	6×10^{18}	1×10^{19}	7×10^{18}
n_i (cm^{-3})	2.3×10^{13}	1×10^{10}	2.1×10^{6}
m_e^*/m_e	0.88	1.08	0.067
m_h^*/m_e	0.29	0.55	0.47
b_e ($cm^2\ V^{-1}s^{-1}$)	3800	1450	8500
b_h ($cm^2\ V^{-1}s^{-1}$)	1800	480	400

Standard Global *AM*1.5 Spectrum

with 1000 W m^{-2} total radiation

λ (μm)	$dj_E/d\lambda$ ($W\ m^{-2}\ \mu m^{-1}$)	λ (μm)	$dj_E/d\lambda$ ($W\ m^{-2}\ \mu m^{-1}$)	λ (μm)	$dj_E/d\lambda$ ($W\ m^{-2}\ \mu m^{-1}$)
0.3050	9.5	0.7400	1271.2	1.5200	262.6
0.3100	42.3	0.7525	1193.9	1.5390	274.2
0.3150	107.8	0.7575	1175.5	1.5580	275.0
0.3200	181.0	0.7625	643.1	1.5780	244.6
0.3250	246.8	0.7675	1030.7	1.5920	247.4
0.3300	395.3	0.7800	1131.1	1.6100	228.7
0.3350	390.1	0.8000	1081.6	1.6300	244.5
0.3400	435.3	0.8160	849.2	1.6460	234.8
0.3450	438.9	0.8237	785.0	1.6780	220.5
0.3500	483.7	0.8315	916.4	1.7400	171.5
0.3600	520.3	0.8400	959.9	1.8000	30.7
0.3700	666.2	0.8600	978.9	1.8600	2.0
0.3800	712.5	0.8800	933.2	1.9200	1.2
0.3900	720.7	0.9050	748.5	1.9600	21.2
0.4000	1013.1	0.9150	667.5	1.9850	91.1
0.4100	1158.2	0.9250	690.3	2.0050	26.8
0.4200	1184.0	0.9300	403.6	2.0350	99.5

λ (μm)	$dj_E/d\lambda$ ($W\,m^{-2}\,\mu m^{-1}$)	λ (μm)	$dj_E/d\lambda$ ($W\,m^{-2}\,\mu m^{-1}$)	λ (μm)	$dj_E/d\lambda$ ($W\,m^{-2}\,\mu m^{-1}$)
0.4300	1071.9	0.9370	258.3	2.0650	60.4
0.4400	1302.0	0.9480	313.6	2.1000	89.1
0.4500	1526.0	0.9650	526.8	2.1480	82.2
0.4600	1599.6	0.9800	646.4	2.1980	71.5
0.4700	1581.0	0.9935	746.8	2.2700	70.2
0.4800	1628.3	1.0400	690.5	2.3600	62.0
0.4900	1539.2	1.0700	637.5	2.4500	21.2
0.5000	1548.7	1.1000	412.6	2.4940	18.5
0.5100	1586.5	1.1200	108.9	2.5370	3.2
0.5200	1484.9	1.1300	189.1	2.9410	4.4
0.5300	1572.4	1.1370	132.2	2.9730	7.6
0.5400	1550.7	1.1610	339.0	3.0050	6.5
0.5500	1561.5	1.1800	460.0	3.0560	3.2
0.5700	1507.5	1.2000	423.6	3.1320	5.4
0.5900	1395.5	1.2350	480.5	3.1560	19.4
0.6100	1485.3	1.2900	413.1	3.2040	1.3
0.6300	1434.1	1.3200	250.2	3.2450	3.2
0.6500	1419.9	1.3500	32.5	3.3170	13.1
0.6700	1392.3	1.3950	1.6	3.3440	3.2
0.6900	1130.0	1.4425	55.7	3.4500	13.3
0.7100	1316.7	1.4625	105.1	3.5730	11.9
0.7180	1010.3	1.4770	105.5	3.7650	9.8
0.7244	1043.2	1.4970	182.1	4.0450	7.5

References

1 Falk, G. and Ruppel, W. **(1976)** *Energie und Entropie*, Springer-Verlag, Berlin.
2 Würfel, P. **(1982)** *Journal of Physics C*, **15**, 3967.
3 (a) Shockley, W. and Read, W.T. **(1952)** *Physical Review*, **87**, 835; (b) Hall, R.N. **(1951)** *Physical Review*, **83**, 228.
4 Würfel, P. Finkbeiner, S. Daub, E. **(1995)** *Appl. Phys. A*, **60**, 67.
5 Landau, L.D. and Lifschitz, E.M. **(1980)** *Course of Theoretical Physics*, Vol. V, Butterworth-Heinemann.
6 Kittel, C. **(1958)** *Elementary Statistical Physics*, John Wiley & Sons, Ltd.
7 deVos, A. **(1983)** *Proceedings of the 5th E.C. Photovoltaic Solar Energy Conference*, Athens, p. 186.
8 O'Reagan, B. and Grätzel, M. **(1991)** *Nature*, **353**, 737.
9 Jaeger, K. and Hezel, R. **(1985)** *Proceedings of the 18th IEEE PV Specialists Conference*, Las Vegas, p. 388.
10 Tang, C. **(1986)** *Applied Physics Letters*, **48**, 183.
11 Walzer, K., Maennig, B., Pfeiffer, M., and Leo, K. **(2007)** *Chemical Reviews*, **107**, 1233.
12 Shaheen, S.E., Brabec, C.J., Padinger, F., Fromherz, T., Hummelen, J.C., and Sariciftci, N.S. **(2001)** *Applied Physics Letters*, **78**, 841.
13 Organic Photovoltaics, ed. by C. Brabec, O. Scherf and V. Dyakonov **(2008)**, Wiley-VCH.
14 Yablonovitch, E. **(1982)** *Journal of the Optical Society of America*, **72**, 899.
15 Zhao, J., Wang, A., Green, M., and Ferrazza, F. **(1998)** *Applied Physics Letters*, **73**, 1991.
16 Brabec, C.J., Sariciftci, N.S., and Hummelen, J.C. **(2001)** *Advanced Functional Materials*, **11**, 15.
17 Green, M.A. **(2003)** *Third Generation Photovoltaics*, Springer-Verlag.
18 Shockley, W. and Queisser, H.J. **(1961)** *Journal of Applied Physics*, **32**, 510.
19 Trupke, T. and Würfel, P. **(2004)** *Journal of Applied Physics*, **96**, 2347.
20 Sinton, R.A., Verlinden, P.J., Crane, D.E., Swanson, R.M. **(1988)** *Proceedings of the 8th E.C. Photovoltaic Solar Energy Conference*, Florence, p. 1472.
21 King, R.R., Law, D.C., Edmondson, K.M., Fetzer, C.M., Kinsey, G.S., Yoon, H., Sherif, R.A., Krut, D.D., Ermer, J.H., Hebert, P., Peichen, P. and Karam, N.H., **(2007)** *Proceedings of the 22nd European Photovoltaic Solar Energy Conference*, Milan, p.11.
22 Green, M.A. **(2000)** *Proceedings of the 16th European Photovoltaic Solar Energy Conference*, Glasgow, p. 51.
23 Werner, J.H., Brendel, R., and Queisser, H.J. **(1994)** *First World Conference on Photovoltaic Energy Conversion*, Hawaii, p.1742.
24 Würfel, P. **(1996)** *Solar Energy Materials and Solar Cells*, **46**, 43.
25 Ross, R.T. and Nozik, A.J. **(1982)** *Journal of Applied Physics*, **53**, 3813.

Physics of Solar Cells: From Basic Principles to Advanced Concepts. Peter Würfel

ISBN: 978-3-527-40857-3

26 Luque, A. and Marti, A. (**1997**) *Physical Review Letters*, **78**, 5014.
27 Trupke, T., Green, M.A., and Würfel, P. (**2002**) *Journal of Applied Physics*, **92**, 4117.
28 Trupke, T., Green, M.A., and Würfel, P. (**2002**) *Journal of Applied Physics*, **92**, 1668.

Index

Physics of Solar Cells: From Basic Principles to Advanced Concepts. Peter Würfel

ISBN: 978-3-527-40857-3